ACTIN
Biophysics, Biochemistry, and Cell Biology

Recent Volumes in this Series

ACTIN
Biophysics, Biochemistry, and Cell Biology

Edited by

James E. Estes
VA Medical Center
Albany, New York

and

Paul J. Higgins
Albany Medical College
Albany, New York

PLENUM PRESS • NEW YORK AND LONDON

Library of Congress Cataloging-in-Publication Data

Actin : biophysics, biochemistry, and cell biology / edited by James
  E. Estes and Paul J. Higgins.
       p.   cm. -- (Advances in experimental medicine and biology ; v.
  358)
    "Proceedings of an International Conference on the Biophysics,
  Biochemistry, and Cell Biology of Actin, held August 5-9, 1992, in
  Troy, New York"-- T.p. verso.
    Includes bibliographical references and index.
    ISBN 0-306-44810-6
    1. Actin--Congresses.   2. Cytoplasmic filaments--Congresses.
  I. Estes, James E.   II. Higgins, Paul J.   III. International
  Conference on the Biophysics, Biochemistry, and Cell Biology of
  Actin (1992 : Troy, N.Y.)   IV. Series.
  QP552.A27A197   1994
  591.1'852--dc20   94-26868
                                                                  CIP

Proceedings of an International Conference on the Biophysics, Biochemistry, and Cell Biology of Actin,
held August 5-9, 1992, in Troy, New York

ISBN 0-306-44810-6

©1994 Plenum Press, New York
A Division of Plenum Publishing Corporation
233 Spring Street, New York, N.Y. 10013

Printed in the United States of America

PREFACE

During the period August 5-9, 1992, and immediately preceding the 1992 Gordon Research Conference on Motile and Contractile Systems, the "Third International Conference on the Structure and Function of Ubiquitous Cellular Protein Actin" was held at the Emma Willard School in Troy, New York, under the title "ACTIN '92". This conference focused on the fundamental properties and cellular functions of actin and actin-based microfilament systems. The first conference in this series was held in 1982, in Sydney, Australia, and hosted by Dr. Cristobal G. dos Remedios and Dr. Julian A. Barden, both from the University of Sydney (New South Wales, Austrailia). The second conference convened in Monza, Italy in June 1987, and was organized by Dr. Roberto Colombo, University of Milan (Italy). This third gathering of researchers devoted to the study of actin and actin-associated proteins was organized by Dr. James E. Estes, Albany Stratton VA Medical Center and Dr. Paul J. Higgins, Albany Medical College, who were assisted by an Organizing Committee consisting of Dr. Edward D. Korn (National Heart, Lung and Blood Institute, NIH), Dr. Thomas P. Stossel (Massachusetts General Hospital), Dr. Fumio Matsumura (Rutgers University), and Dr. Stephen Farmer (Boston University). This meeting was dedicated to the many pioneering contributions of Professor Fumio Oosawa to the field of actin research.

Written versions of oral presentations by the invited speakers, and of selected poster contributions, are presented here in essentially the same order as the meeting format which focussed on: 1) basic properties of the actin molecule and actin-based microfilament systems, 2) actin-associated proteins and control of filament-based assembly-disassembly, and 3) cellular functions of the microfilament system.

The Organizing Committee for ACTIN '92 is pleased to acknowledge the financial support and helpful assistance of: the National Institutes of Health (for NIH grant 1 R13 AR41437-01), the Albany Stratton VA Medical Center (including staff from Research Service, Education Service, Engineering Service, and Medical Media), Albany Medical College, Matsushita Electric Industrial Company, East Greenbrush Travel Agency, and ler Graphics. During the course of the three-day conference, several individuals were particularly instrumental in its success, and the Organizing Committee for ACTIN '92 wishes to specifically acknowledge the contributions of: Dean Anthony Tartaglia, Dr. A. Daoud, Lynn Selden, Henry Kinosian, Michael Ryan, Chris Keenan, John Flor, Axel Herrmannsdoefer, Jack Hemenway, Millie Estes, Denise Higgins and Joanna Estes. Finally, we wish to extend a special thank you to Ms. Trudy Hammer and the staff at the Emma Willard School for allowing us the use of their gracious facilities and providing such charming accomodations.

<table>
<tr><td>James E. Estes</td><td>Paul J. Higgins</td></tr>
<tr><td>Stratton VA Medical Center</td><td>Albany Medical College</td></tr>
<tr><td>Albany, New York</td><td>Albany, New York</td></tr>
</table>

CONTENTS

BASIC PROPERTIES OF THE ACTIN MOLECULE AND ACTIN-BASED MICROFILAMENT SYSTEMS

ACTIN-ASSOCIATED PROTEINS AND CONTROL OF FILAMENT-BASED ASSEMBLY-DISASSEMBLY

"

CELLULAR FUNCTIONS OF THE MICROFILAMENT SYSTEM

BASIC PROPERTIES OF THE ACTIN MOLECULE AND ACTIN-BASED MICROFILAMENT SYSTEMS

VIBRATIONAL MODES OF G-ACTIN

Monique M. Tirion,[1] Daniel ben-Avraham,[1] and Kenneth C. Holmes[2]

[1]Clarkson University
Physics Department
Potsdam, NY 13699-5820
[2]Max-Planck-Institute for Medical Research
Jahnstrasse 29
6900 Heidelberg, Germany

INTRODUCTION

The determination of the atomic structure of g-actin (Kabsch et al., 1990, see
Fig. 1) allowed the development of an atomic model for f-actin (Holmes et al., 1990).
The structure of f-actin was deduced from x-ray diffraction patterns from bundles of
aligned actin filaments, using the known helical symmetry of the filament and keeping
the atomic structure of the monomer fixed. The model of f-actin was obtained,
therefore, using only four structural parameters: three rotational and one radial
degree of freedom. The solution thus obtained is unique, and achieved an R-factor of
.22; an extremely good fit with such few parameters.

We expect, however, that the structure of the monomer is modified as it is incorporated into the filament. In an attempt to model structural modifications, we
refined each of the domains and subdomains independently as rigid bodies. This
work made apparent that very minor structural modifications improve the fit to the
x-ray data dramatically, an encouraging indication that the monomer structure is not
distorted substantially as it is incorporated into the filament. However, by modeling
the structural changes using rigid-body refinements we could not maintain proper
stereochemistry: after refinement the various domains and/or subdomains were no
longer continuous, since the polypeptide chain was severed to permit the rigid-body
refinements. Hence we searched for a technique that would permit a refinement algorithm to explore a large portion of phase-space using few structural parameters,
while maintaining proper stereochemistry. These constraints motivated us to study
the normal modes of the monomer, g-actin, in order to model the structural modifications of the monomer as it is incorporated into the filament, and to characterize the
flexibility inherent in this molecule. Here we describe the technique used to determine
the normal modes of the large ternary system, g-actin-ADP-Ca^{++} , as well as the
computed slow modes.

Actin: Biophysics, Biochemistry, and Cell Biology
Edited by J.E. Estes and P.J. Higgins, Plenum Press, New York, 1994

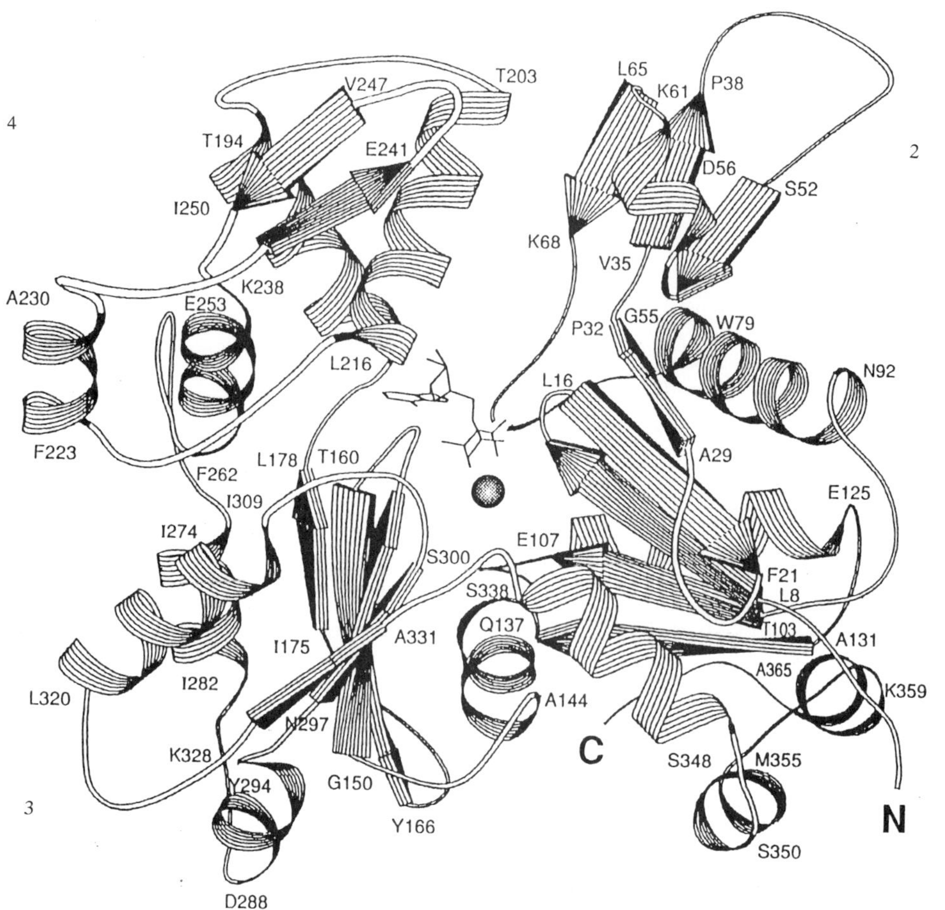

Figure 1. Schematic representation of the three-dimensional structure of actin. First and last amino acid residues in the helices and sheet strands are specified. ADP and Ca^{++} are located between the small (right) and large (left) domains. The small domain is divided into subdomains 1 (1-32, 70-144 and 338-375) and 2 (33-69) and the large domain is subdivided into subdomains 3 (145-180 and 270-337) and 4 (181-269). From Kabsch et al., 1990.

TECHNIQUE

We used standard classical mechanics theory to obtain the normal modes of actin (Goldstein, 1950). Here we follow the protocol and notation of Levitt et al., (1985), as described in Tirion and ben-Avraham (1993). Given a potential energy function, E_p, one minimizes it with respect to the generalized coordinates, q_i, and writes it as a quadratic expansion around this minimum (which is at q_i^0):

$$E_p = \frac{1}{2} \sum_{i,j}^{n} (q_i - q_i^0) F_{ij} (q_j - q_j^0) = \frac{1}{2}(\mathbf{q} - \mathbf{q^0})^\dagger \mathbf{F}(\mathbf{q} - \mathbf{q^0}), \tag{1}$$

where the $\mathbf{F}$ elements are

$$F_{ij} = \left(\frac{\partial^2 E_p}{\partial q_i \partial q_j} \right)_{q_i = q_i^0,\ q_j = q_j^0}. \tag{2}$$

The kinetic energy is written as a quadratic polynomial of the generalized velocities, $\dot{q}_i$ (the dot represents differentiation with respect to time):

$$E_k = \frac{1}{2} \dot{\mathbf{q}}^\dagger \mathbf{H} \dot{\mathbf{q}}, \tag{3}$$

with

$$H_{ij} = \sum_{l}^{N} m_l \frac{\partial \mathbf{r}_l}{\partial q_i} \cdot \frac{\partial \mathbf{r}_l}{\partial q_j}, \tag{4}$$

where the index l runs over all the atoms in the molecule. The $\mathbf{H}$ matrix also allows the transformation from Cartesian coordinates $\mathbf{r}$ to generalized coordinates q, when all m_l elements are set to unity. The derivatives with respect to q are moving derivatives in which overall translation and rotation of the molecule must be eliminated.

The modes are obtained from the solution to the equation

$$\mathbf{FA} = \mathbf{\Lambda HA}, \tag{5}$$

subject to the normalization condition

$$\mathbf{A}^\dagger \mathbf{HA} = \mathbf{I}. \tag{6}$$

The eigenfrequencies are then given by the elements of the diagonal matrix $\mathbf{\Lambda}$, $\omega_k^2 = \Lambda_{kk}$, and the eigenvectors are the columns of the matrix $\mathbf{A}$. A general motion of the molecule can be expressed in terms of the normal modes:

$$q_j = q_j^0 + \sum_{k}^{n} A_{jk} \alpha_k \cos(\omega_k t + \delta_k). \tag{7}$$

The normalization condition, Eq. (6), ensures that both the potential and the kinetic energy can now be written as a sum of pure squares of $\alpha_k \cos(\omega_k t + \delta_k)$ and their time-derivatives (*i.e.*, the Hamiltonian is diagonalized). The normal modes and eigenfrequencies can be employed in a standard manner to yield a variety of interesting parameters, including the temperature factor, r.m.s. fluctuations of different atoms and correlations of motion.

Dynamic Variables

We analyzed the system of monomeric actin bound with ADP and Ca^{++}. The crystal coordinates consisted of 372 residues, the last 3 carboxy-terminal residues not being defined by the crystallographic data. In order to reduce the total number of structural parameters of this 3539 atom system to a manageable level, we chose as our dynamic variables (*i.e.*, generalized q coordinates) only single bond torsions (1384 degrees of freedom in all). Normal mode studies done on bovine pancreatic trypsin inhibitor (BPTI) using bond, angle and torsion degrees of freedom provide very similar results for the softest, slowest modes as studies that include only the single bond torsions (Brooks and Karplus, 1983; Levitt et al., 1985).

Energy Parametrization

We used a potential energy function, L79, derived by Levitt (1983). It includes separate expressions for torsion energies, as well as non-bonded, Lennard-Jones type van der Waal's energies. The non-bonded energy terms included all atom pairs separated by less than the sum of their van der Waal's radii plus an additional 2 Å, and separated by more than 3 consecutive bonds (since the distance of atoms separated by less than four bonds cannot be changed by torsion angle changes). Hydrogen bonds were parametrized using directional hydrogen bonds. Only hydrogen atoms available for hydrogen bond interactions were included. The NH, NH_2 and NH_3 hydrogens on lysines and arginines were excluded to avoid these long side chains from folding back onto the protein surface, which they do in the absence of solvent. Otherwise, the analysis was done in vacuo and other solvent effects were disregarded.

RESULTS

The analysis provides 1384 normal modes (as many modes as there are degrees of freedom) for the actin system. Each mode describes a "natural" mode or oscillation of the actin system about the single-bond torsion degrees of freedom. Each mode is coherent; it maintains a constant shape or pattern during its oscillation, analogues to the sinusoidal oscillations of a violin string. By summing the modes, with suitable relative amplitudes and phases, it should then be possible to describe any internal motion of the monomer due to torsional variations. Insofar as the molecular potential energy surface is harmonic, or quadratic, about the mimimum, this is an accurate description of the motion. The harmonic approximation becomes more accurate for small activation energies, or amplitudes, of each mode. (However, it is possible that only small adjustments in the orientations of a few groups of atoms during an oscillation, adjustments not predicted by the normal mode algorithm, would "accomodate" the mode in an energetically favorable manner over larger amplitudes.)

The frequencies, or periods, of the modes extend from a slow of 17 psec to a fast of 0.1 psec. The fastest modes are associated with rapid oscillations of small groups of atoms, such as side-chains on the surface of the molecule, and do not exhibit long-range correlations among the various degrees of freedom. The slowest computed modes exhibit long-range correlations in the motions of various domains and subdomains, as described in the following sections.

According to classical dynamics, each normal mode will have a time-averaged potential energy of $\frac{1}{2}k_BT$ above the value at the minimum. (k_B is Boltzmann's constant, and T is the absolute temperature). Since it is easy to compute the potential energy of any one mode, it is possible to relate the thermal energy available in a heat bath to the amplitude of activation of each mode. This permits a quantitative

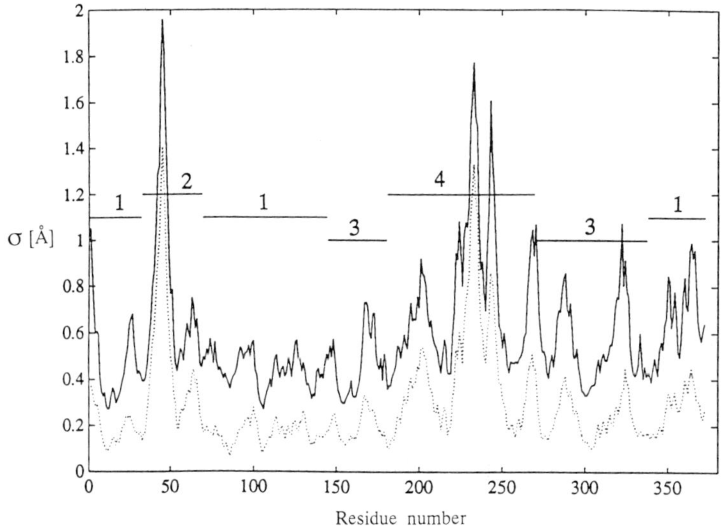

Figure 2. The r.m.s. fluctuations due to the combined effect of the first 4 modes, $\sigma_{\alpha i}^{1 to 4}$ (dotted line), and due to all modes, $\sigma_{\alpha i}$ (solid line), as a function of residue number. The first four modes are seen to contribute above 50% to the total rms deviation of the α-carbons at room temperature.

description of $\sigma_{C_{\alpha i}}^{k}$: the rms deviation of the ith C_α-carbon atom due to mode k, and $\sigma_{C_{\alpha i}}$: the rms deviation of the ith C_α-carbon atom due to *all* modes.

Fig. 2 shows a plot of the rms deviation of each C_α-carbon atom due to all modes, $\sigma_{C_{\alpha i}}$, at $T = 300K$ (solid line), and the rms deviation of each C_α-carbon atom due to only the slowest four modes $\sigma_{C_{\alpha i}}^{k=1,4}$ (dashed line). It is found that over 50% of the overall motion associated with the monomer at room temperature can be described by the first four, slow modes. (Over 72% of the motion can be described by the 12 slowest modes). It is for this reason that we concentrate our attention on the motions associated with these slowest modes.

Temperature Factors

It is possible to compute theoretical temperature factors, B, for each atom using the normal modes. Fig. 3 shows a comparison of the theoretical temperature factors (solid line) and the experimentally determined temperature factors derived from x-ray crystallographic data (dashed line). There is a clear correlation between the two curves, with peaks in the experimental data also appearing in the theoretical curve. One area of significant discrepency is from residues 42-53. This loop, located in subdomain 2 (see Fig. 1) forms a crystal contact with the enzyme, DNase I, which was co-crystallized with g-actin. Residues 42-44 of actin form one strand of a β-strand, the other strand of the two-stranded β-sheet being contributed by the DNase

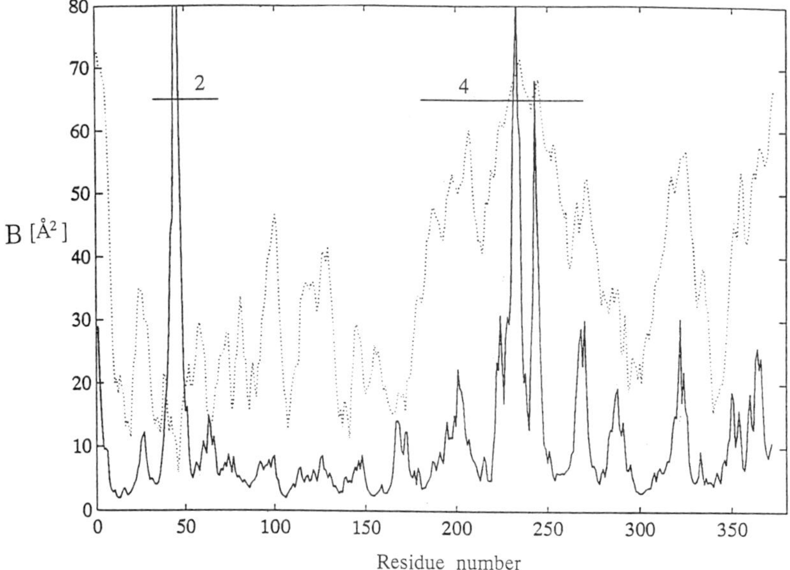

Figure 3. Comparison of the experimental (dotted line) and calculated (solid line) temperature factors of the α-carbon atoms. The ith calculated B value is derived from the r.m.s. fluctuation of the ith α-carbon atom when all the modes are excited at 300K. Subdomains 2 and 4 are seen to have the largest B factors and are indicated in the figure.

I. In the absence of the co-enzyme, this loop is unconstrained and free to vibrate, as the data shows. Furthermore, it is seen that the experimentally observed temperature factors have larger magnitudes than the computed ones. This is due to the fact that the computed B values consider motion of atoms strictly due to internal vibrations of the protein. The experimentally determined B values record not just the mobility due to internal vibrations, but also record crystal lattice vibrations and external noise. In fact, the normal mode analysis can be used in order to explore crystal lattice packing constraints, as well as to separate internal from external vibrations and motions (see, for example, Diamond, 1990).

Mode 1: Propeller Motion

The slowest vibrational mode of g-actin, with a computed frequency of 17 psec, we have termed the propeller mode. It pertains to a twisting of the small and large domains about the ADP binding loops (residues 14-16 and 157-159). The hinge points that permit the relative reorientation of the two domains are located at gln-137 and val-339, as can be verified by the small $\sigma^1_{C_\alpha}$ values of these residues in Fig. 4a.

During this motion, subdomains 1 and 2 move in unison, with subdomain 1 coming out of the plane of the page as subdomain 2 moves into the plane of the page, and vice versa, as seen in the orientation of Fig. 1. Subdomains 3 and 4 perform a similar motion, but antiparallel to the small domain: subdomain 4 moves out of the plane of the page as 2 moves in.

The gln-137 hinge-point is the last residue of the $\beta\alpha\beta$ motif of subdomain 1, and the first residue of the α-helix at 137-144; this pivot point lies at the junction of two elements of secondary structure, with no intervening residues. The polar side chain of gln-137 is oriented towards the nucleotide binding site, such that its $O_{\epsilon 1}$ is 3.5 Å removed from the Ca^{++}. Val-339 is located at the start of the α-helix directly above this, and marks the start of the C-terminal region of the small domain. The 137-144 helix is situated between the large and small domains, just below the nucleotide binding cleft, with its axis running parallel to the plane of the β-sheet in subdomain 3.

Another characteristic movement of this mode pertains to residues 223-250 in the large domain. Residues 216-222 in subdomain 4 move in unison with the large domain, while residues 230-250 swing in and out of the plane of the paper. This has the effect of revolving the intervening 223-230 α-helix as a rigid body around its edge at residue 223. This motion is seen to correspond to the large peak in Fig. 4a, and also contributes to the large calculated B factor of this region.

Mode 2: Rolling Motion

From Fig. 4b, it is seen that for the next slowest mode the small subdomains (2 and 4) are more mobile than the larger subdomains (1 and 3). Residues 182-252 in subdomain 4 roll around the stationary α-helix 253-262, whereby residues 219 and 237 remain nearly stationary, as they pass near to this helix. It is interesting to note that the global minimum of Fig. 4b occurs at residue 258, which is a proline in the middle of the 253-262 helix. Also, the loop at residues 262-274 swings in and out with the rest of subdomain 4. At the same time, residues 34-55 in subdomain 2 swing into and out of the plane of the figure, with the DNase I binding loop moving the most. Residues 70-77, which form the loop at the back of the nucleotide- binding site, remain fairly steady.

Mode 3: Scissors Motion

The third slowest mode, with a period of 12 psec, has the appearance of a scissor-motion: the small and large domains move independently in the plane of Fig. 1 to alternately open and close the clefts at the top and bottom of the protein. The hinges separating the relative motion of the large and small domains are at residues ala-331 and at helix 137-144, as seen by their small $\sigma^4_{C\alpha i}$ values in Fig. 4d. The amino-terminal end of the helix, gln-137, moves in tandem with the small domain, while the other end of the helix, ala-144, moves in tandem with the large domain. This helix, then, acts like a classical "oily" spring; consisting almost exclusively of small, hydrophobic residues, it permits motion of the two domains perpendicular to the helix axis. The other hinge point, ala-331, is the last residue of the $\beta\alpha\beta$ motif in subdomain 3. Ala-331 is followed by two prolines at 332-333 leading back to the carboxy terminus in subdomain 1. These two prolines, breaking the pattern of hydrogen bonding, are ideally situated to permit a discontinuity in the motion.

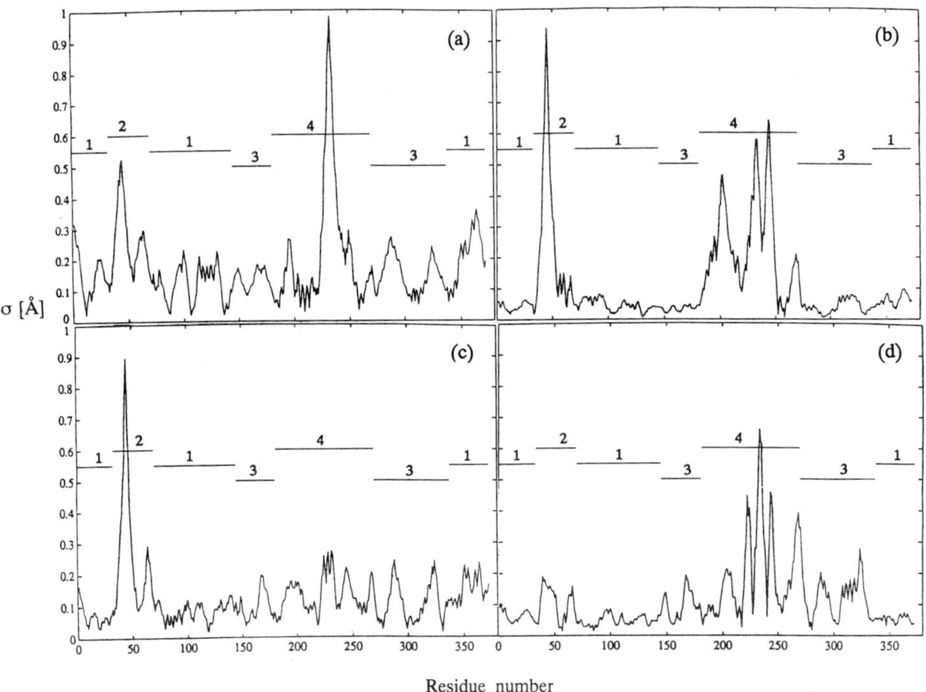

Figure 4. The variation with residue number, i, of the r.m.s. fluctuation of each α-carbon atom, $\sigma^k_{\alpha i}$, for each of the 4 slowest frequency modes ($k=1$ to 4 in (a) to (d) respectively). The 4 subdomains are indicated with the labelled horizontal bars.

Mode 4 and Higher Modes

Figure 4d shows that for next slowest mode residues in the large domain have large amplitudes of vibration. And indeed, when viewed on the graphics display, we see that subdomains 3 and 4 appear to 'hug' forward toward each other with the result that the loop at residues 262-274 between them extends further out towards the back. Also, the carboxy-terminal region in subdomain 1, being 'draped' over this subdomain like an arm, vibrates slightly, in tandem with subdomain 3.

As the frequency of the normal modes increases, the wavelength of the collective motion decreases. Thus, rather than a coherent motion of whole subdomains, the motion is broken into smaller components. In mode 5 one can still identify coherent motion of subdomain 1 which twists slightly around, toward subdomain 3. The rest of the molecule's motion becomes hard to describe due to the small coherence length.

DISCUSSION

The normal mode analysis of g-actin provides insight into the inherent flexibility of this monomer, as well as insight into the general properties of vibrational spectra of globular proteins (ben-Avraham, 1993). The analysis provides theoretical evidence for large-scale, collective motions of domains and subdomains, evidence borne out by the correlation between the computed and observed temperature factors. The analysis identifies hinge, or pivot, points in the protein: main-chain residues whose orientation breaks the pattern of hydrogen bonding to permit relative reorientation of various domains. One of these hinge points, gln-137, is located at the junction of a β-strand and an α-helix seperating the large and small domains of actin. Its polar side-chain is oriented towards the nucleotide-binding site, as if to "sense" the state of the bound nucleotide.

Another hinge seems to be provided by the 137-144 helix separating the small and large domains. It is composed almost entirely of small nonpolar residues (ala-val-leu-ser-leu-tyr-ala) that allow this helix to behave somewhat as a classical "oily" spring that lacks long side-chains that could get entangled with surrounding residues. It is very interesting to note that a similar structural motif is seen in both hexokinase and the ATPase fragment of a 70kD heat shock cognate proteins (Steitz et al., 1981; Flaherty et al. 1990). Both hexokinase and HSC70 are composed of two equal-sized domains that bind ADP (or ATP) in a cleft deep between the two domains. In each case the two domains are linked by a short helix similar to the one found in actin. Even though there is very low sequence homology between the various proteins, gln-137 is preserved in hexokinase (glu-175) and HSC70 (asp-189), and in each case this residues is found at the junction of an α-helix and a β-strand as in actin. And in each case this residue is oriented so as to be able to interact with the bound nucleotide. It is therefore interesting to hypothesize whether these vastly divergent proteins have this structural similarity in order to maintain similar dynamical flexiblities that permit these proteins to execute their various enzymatic and catalytic activities.

The normal modes were used to refine the f-actin model. Rigid body refinements failed to maintain proper stereochemistry, and cartesian refinement of individual amino acid or atoms exceeded the resolution of the fiber diffraction data. Using instead the normal modes as refinement parameters, we were able to reduce the (fiber diffraction) R-factor from 0.22 to 0.11 using merely 9 degrees of freedom (the slowest 9 modes). The refinement brought the DNase I binding loop to a lower radius and to a higher axial location, in agreement with electron microscopy and cross-linking experiments that do not show density at this high radius, and show that gln-41 is able to cross-link with lys-113 of the subunit above it (Orlova and Egelman, 1993; Hegyi et al., 1992). As it is very likely that the monomer does undergo some structural modifications as it is incorporated into the filamentous form of actin, g-actin, our success in lowering the R-factor significantly using few degrees of freedom indicates that these modes may help identify pathways by which the monomer may be deformed as it is incorporated into the fiber.

We are continuing our studies of f-actin by computing the modes associated with the fiber, using the modes as internal degrees of freedom. This analysis may shed light on the flexibility inherent to the fiber; quantify the apparent angular disorder, the amount of lateral slipping, and the role of internal hinges in providing the observed flexiblity of the filament.

REFERENCES

ben-Avraham, D., 1993, Vibrational normal-mode spectrum of globular proteins, *Phys. Rev. B* 47:14559.

Brooks, B & Karplus, B., 1983, Harmonic dynamics of proteins: Normal modes and fluctuations in bovine pancreatic trypsin inhibitor, *Biophysics* 80:6571.

Diamond, R., 1990, On the use of normal modes in thermal parameter refinement: theory and application to bovine pancreatic trypsin inhibitor, *Acta Cryst.* A46:425.

Egelman, E. H.,Francis, N. & DeRosier, D. J., 1982, F-actin is a helix with a random variable twist, *Nature* 298:131.

Flaherty, K. M.,McKay, D.,Kabsch, W, & Holmes, K. C., 1991, Three-dimensional structure of the ATPase fragment of a 70kD heat-shock cognate protein, *Proc. Natl. Acad. Sci. USA* 88:5041.

Goldstein, H., 1950, "Classical Mechanics", Addison-Wesley, Reading, Massachusetts.

Hegyi, G., Michel, H., Shabanowitz, J., Hunt, D. F., Chatterjee, N., Healy-Louie, G. and Elzinga, M., 1992, Gln-41 is intermolecularly cross-linked to lys-113 in f-actin by N-(4-azidobenzoyl)-putresine. *Protein Science*, 1:132.

Holmes, K. C.,Popp, D.,Gebhard, W,Kabsch, W., 1990, Atomic model of the actin filament, *Nature* 347:44.

Kabsch, W.,Mannherz, H., G.,Suck, D.,Pai, E.,Holmes, K. C., 1990, Atomic structure of the actin:DNase I complex, *Nature* 347:37.

Levitt, M., 1983, Molecular dynamics of native protein. I. Computer simulation of trajectories, *J. Mol. Biol.* 168:595.

Levitt, M.,Sander, C. and Stern, P. S., 1985, Protein normal-mode dynamics: trypsin inhibitor, crambin, ribonuclease and lysozyme, *J. Mol. Biol.* 181:423.

Millonig, R.,Sütterlin, R.,Engel, A.,Pollard, T. D.,Aebi, U., 1989, The 'lateral slipping' model of F-actin filaments, *in:* "Springer Series In Biophysics Vol. 3, Cytoskeletal and Extracellular Proteins" U. Aebi and A. Engel, eds., Springer- Verlag, Heidelberg.

Orlova, A. and Egelman, E. H. , 1993, A conformational change in the actin subunit can change the flexibility of the actin filament.*J. Mol. Biol.* In press.

Steitz, T. A.,Anderson, W. F.,Fletterick, R. J. & Anderson, C. M., 1977, High resolution crystal structures of yeasts hexokinase complexes with substrates, activators, and inhibitors. Evidence of an allosteric control site, *J. Biol.Chem.* 252:4494.

Tirion, M. M. & ben-Avraham, D., 1993, Normal mode analysis of g-actin, *J. Mol. Biol.* 230:186.

COMBINING ELECTRON MICROSCOPY AND X-RAY CRYSTALLOGRAPHY DATA TO STUDY THE STRUCTURE OF F-ACTIN AND ITS IMPLICATIONS FOR THIN-FILAMENT REGULATION IN MUSCLE

Robert Mendelson[1] and Edward Morris[2]

[1]Dept. of Biochemistry and Biophysics
  and Cardiovascular Research Institute
 University of California, San Francisco, CA 94143

[2]Biophysics Section
  Blackett Laboratory
  Imperial College London

INTRODUCTION

Actin filaments (F-actin) are found in nearly all eukaryotic cells as elements of the cytoskeleton. They also play a central role in various types of contractility, motility and transport. F-actin is a helical polymer composed of identical globular subunits, each of which contains 375 amino acids. The atomic structure of the monomer (G-actin; 42 kD) has recently been determined from a complex of the monomer and DNase (Kabsch *et al.*, 1990). The monomer structure, shown in figure 1, has two major domains (historically these were termed "large" and "small", but it is now known that they are of nearly the same size) which are each divided into two subdomains. Subdomain 1 contains the N- and C-termini of the polypeptide chain. The prominent cleft between the two major domains is the site of nucleotide binding. Knowledge of the precise arrangement of the actin subunits within F-actin would be helpful in understanding the function of F-actin at the molecular level.

Holmes *et al.* (1990) investigated the orientation of the actin subunits within the filament by fitting the X-ray fiber diffraction pattern from oriented gels of F-actin complexed with phalloidin using the crystallographic structure of Kabsch *et al.* (1990). This fitting was achieved by computing model intensities obtained after rotations of the G-actin structure and subsequent radial positioning based on the cross-sectional radius-of-gyration (R_c) from X-

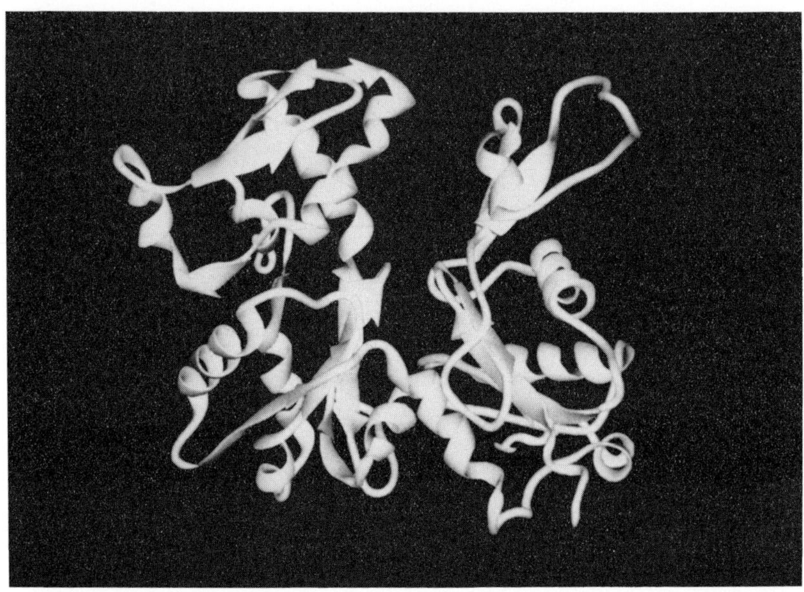

Figure 1. A ribbon representation of the actin monomer from actin-DNase I crystallography by Kabsch *et al.*, 1990. By convention, the "small" domain is divided into subdomain 1 (lower right; residues 1-32, 70-144 and 338-375) and subdomain 2 (upper right; 33-69). The "large" domain is divided into subdomain 3 (lower left; 145-180 and 270-337) and subdomain 4 (upper left; 181-269). ADP and Ca^{2+} (not shown) are located in the cleft between the large and small domains. DNase I binds to the large loop in subdomain 2 that protrudes towards the upper right. Figure generated by RIBBON subprogram of MIDAS (Ferrin *et al.*, 1988).

ray-solution-scattering measurements[1]. From this analysis one orientation of the subunit agreed with the fibre diffraction data significantly better than any of the alternatives. The model orientation had strong inter-protomer connections along the right-handed long-pitch (2-start) helices and weaker connections along the (1-start) left-handed genetic helix.

Despite the ability of the Holmes *et al.* model to account for a significant body of experimental information, it is not certain that the model is both unique and correct. The F-actin structure has not been "solved" in a rigorous crystallographic sense: the X-ray fiber diffraction results are lacking in phase information and do not extend to atomic resolution. In addition, there are other limitations: the use of F-actin phalloidin to simulate actin; fitting a limited portion of reciprocal space; and the assumption that the monomer structure is unchanged upon incorporation into the filament. Thus it is difficult to assess whether the best-fitting model obtained is unique and whether all other models can be rejected with certainty.

Schutt and his colleagues (1989) proposed a very different orientation for the actin subunit based upon the intermolecular contacts observed in the analysis of actin-profilin ribbons. Here contacts between subunits were postulated to occur only along the genetic helix (in agreement with some electron microscopy results.) This model of F-actin is a key element in a recent hypothesis of the mechanism of contraction by Schutt and Lindberg

---

1. This method was actually devised earlier (Mendelson *et al.*, 1984) to fit electron microscopy data, but errors in the (6 Å resolution) extracted G-actin structure (Suck *et al.*, 1981) inhibited analysis.

A number of electron microscope studies of actin filaments have tended to emphasize the protomer contacts along the genetic helix compared to those along the long-pitch helices (Egelman and DeRosier, 1983; Trinick *et al.*, 1986; Milligan and Flicker, 1987). This view of F-actin readily allowed the subunits to have the substantial cumulative angular disorder that was postulated by Egelman and De Rosier to explain the apparent variations in the long-pitch periodicity (Hanson, 1967) of F-actin. Other electron microscope studies appear to show more substantial links between subunits along the 2-start helix (O'Brien *et al.*, 1983; Milligan *et al.*, 1990; Bremer *et al.*, 1991). Bremer *et al.* argued that the variability in cross-over spacing arises from a lateral slippage of the actin subunits which is associated with weak genetic and strong long-pitch interactions. Despite these differences in interpretation, recent electron microscope studies seem to be more consistent with the subunit orientation in the Holmes *et al.* model (Egelman, 1992; Bremer *et al.*, 1991).

With these matters in mind, we have undertaken a systematic analysis to find which orientation of subunits agrees best with actin filament data derived from the electron microscope. The approach is similar to that of Holmes *et al.* (1990) in that a global search is conducted of all possible orientations with a particular angular step size. All minima are located and the monomer is then "rocked" to obtain the best fitting orientation. In the present case, rather than layer-lines from the X-ray diffraction pattern, we use layer-lines calculated from electron microscope images, so they contain phase as well as amplitude information. Although the data are of lower resolution, they contain additional information, and thus this study is complementary to the analysis using X-ray fiber diffraction data. Here we present the first report of our attempt to analyze a number of independent data sets obtained by a variety of electron microscope techniques with the aim of making an objective assessment of their agreement with the proposed models.

In addition to these studies on F-actin, we point out an implication of our results for the mechanism of regulation of vertebrate muscle contraction.

METHODS

In order to fit the data, a reciprocal-space global search procedure (Mendelson *et al.*, 1984) was used. Rotated Fourier transforms were compared to those of the data by a goodness-of-fit parameter which was either the average amplitude-weighted phase residuals (DeRosier and Moore, 1970)

$$Q \equiv \Sigma \langle |F_k| \rangle |\Delta\varphi_k| / \Sigma \langle |F_k| \rangle \tag{1}$$

or an R-factor defined as

$$R \equiv \Sigma |F_k^{expt} - F_k^{calc}|^2 / \Sigma |F_k^{calc}|^2. \tag{2}$$

Here F_k's are the structure factors; φ_k's are the phases; and the sums are over points on layer-lines that are characteristic of helical Fourier transforms. The data filaments were brought into the best fitting alignment with the model filament for *each* monomer orientation by an azimuthal rotation and a longitudinal translation. All possible Euler-angle combinations were examined in 15° intervals, and after all major minima were located, the monomers were "rocked" into the final best-fitting orientation with final angular step sizes of 2.5° or 1°.

The equatorial portion of the Fourier-Bessel transform is invariably distorted in both unstained cryo-electron microscopy and negatively-stained electron microscopy. Such dis-

unstained cryo-electron microscopy and negatively-stained electron microscopy. Such distortions of this non-helical portion of the transform can arise from flattening, accumulation of stain at the edges of the filament, or other experimental non-idealities. For this reason equatorial information from electron microscopy was not included in the fitting. However the R_c, which determines the lowest resolution (s) portion of the equatorial amplitude, is a strong experimental constraint. We thus examined fits at $R_c = 2.35$ nm and $R_c = 2.5$ nm, which are within the range of the value used for fitting the X-ray F-actin fiber pattern (2.352 nm) and various X-ray solution scattering determinations (Hartt & Mendelson, 1980; Matsudaria *et al.*, 1987.) Another piece of external equatorial information that we used was the phase of the second oscillation. Based on X-ray fiber diffraction results of F-actin and F-actin-tropomyosin (Popp and Holmes, 1992) we constrained our fits to have an average phase of $180 \pm 20°$. Computed and data transforms were normalized by the ratio of the summed (non-equatorial) amplitudes.

For negatively-stained data, in addition to equatorial distortions, there can be variations in stain density along the particle associated with the helical repeat which give rise to strong negative peaks outside the molecular envelope of the particle in helical reconstructions. To obviate this effect the layer-line sets used for fitting are calculated from a projected helical reconstruction from which negative density peaks outside a reasonable molecular envelope have been removed. We have observed a very marked improvement in the goodness-of-fit parameters obtained from layer-line sets "thresholded" in this way.

RESULTS

F-actin Structure

Shown in figure 2 are the results of fitting (1) unstained single actin filaments embedded in vitreous ice (as in Milligan *et al.*, 1990, but with the contrast-transfer function corrections supplied by R. Milligan), (2) negatively-stained single actin filaments (Bremer *et al.*, 1991), and (3) actin filaments obtained from images of acto-tropomyosin single-layer paracrystals (Morris, 1979; O'Brien *et al.*, 1983). Displayed are the parameters of the four protomer orientations yielding the best fits. These orientations are grouped according to their primary connectivity and, within this connectivity class, according to their overall orientation (see caption).

For the cryo-microscopy data, two very different monomer orientations produced best fits that were almost equally good when $R_c = 2.35$ nm. Remarkably, while one orientation (2′) is very similar to the two-start helix connectivity model obtained by Holmes *et al.* (1990), the second orientation (1′) has exclusively one-start connectivity and appears to be similar to that proposed by Schutt *et al.*[2] For $R_c = 2.35$ nm, the R-factor fits, which use both amplitudes and phases, slightly favor the 2′ orientation, while fits using Q, which is an amplitude-weighted measure of phase error, give virtually identical values for the two orientations. Surface views of three-dimensional reconstructions of these two fits and the data are shown in Figure 3a-3c. In the 1′ orientation the narrow dimension of the monomer is aligned roughly with the filament axis, and subdomains 3 and 4 form the low-radius connections between subunits along the genetic helix, with subdomains 1 and 2 forming high-radius projections. A high-resolution representation of these two orientations is shown

2.   In relating the 1′ orientation to the structure proposed by Schutt *et al.* we assume that location of Cys-374 in the high-radius domain of their ribbon structure means that it is made up of subdomains 1 and 2 (that is the small domain) rather than the large domain (subdomains 3 and 4) as stated in their original paper (Schutt *et al.*, 1989).

in figure 4. For $R_c = 2.5$ nm, the phase of the second oscillation on the equator produced by the 1′ orientation shifts to zero so that the 1′ orientation is no longer an acceptable fit. Thus for $R_c = 2.5$ nm the 2′ orientation is the most favored for the Milligan *et al.* data.

Data from other methods do not necessarily distinguish between these two possibilities. Measurements of the radius of Cys374 vary between 2.0-2.5 nm (Kasprzak *et al.*, 1988) and 3.7 nm (Taylor *et al.*, 1981) from fluorescence energy transfer and 2.7 nm from electron

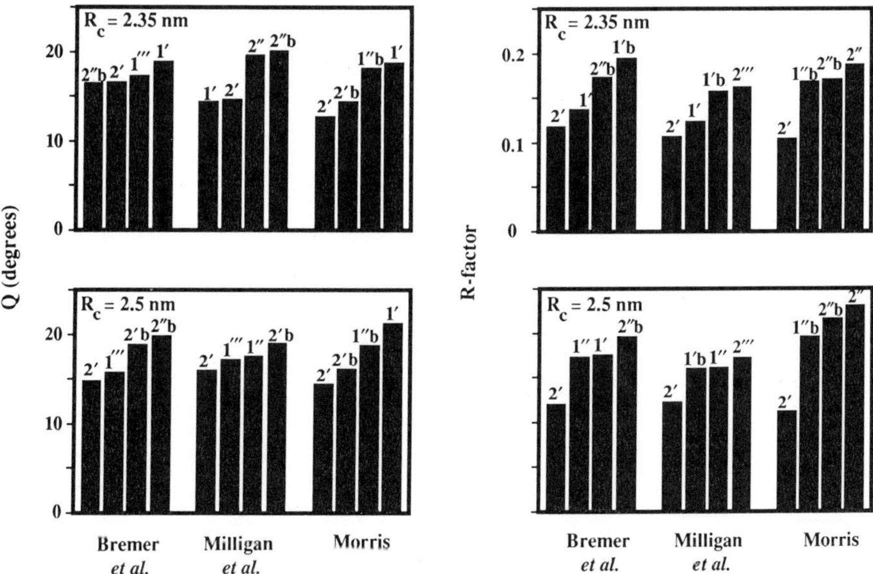

Figure 2. Results of global searches on three electron microscopy data sets. Two goodness-of-fit parameters, the R-factor and the phase residual, are shown for the four best-fitting minima at two plausible cross-sectional radii-of-gyration. The notation used to designate orientations is as follows: The numerals (1 or 2) refer to whether the primary inter-protomer connectivity is along the one-start genetic helix or along the long-pitch two-start helices. The number of primes labels different orientations within the connectivity class. The presence of a trailing "b" indicates that the orientation is symmetry-related to one designated by the same number of primes. Such orientations are related to the original by a 180° rotation about an axis lying in the plane of the monomer between the two major sub-domains. The 1′ and 2′ orientations are similar to Schutt *et al.* and Holmes *et al.*, respectively.

microscopy of SH-directed gold clusters. Based on X-ray diffraction measurements of gold-labeled Cys-374 in actin-profilin ribbons, Schutt *et al.* predicted a radius of 3.3 nm in F-actin. The 1′ and 2′ orientations found here yield radii of 3.1-3.4 and 2.7-2.9 nm, respectively, for $R_c = 2.35$ and 2.5 nm. On the other hand, our calculations show that of the 15 lowest minima examined in the current study, only the 2′ has Cys374-Lys191 and Gln-41-Lys113 distances of less than 2 nm, as would be required by the chemical cross-linking of these residues (El-zinga and Phelan, 1984; Hegyi *et al.*, 1992). This strongly supports the 2′ type of orientation.

One possible source of this near orientation degeneracy is the resolution of the data. The resolving power (s_{max}) of the unstained cryo data was near $1/2.9$ nm^{-1}. Fits to negatively-stained data of Bremer *et al.*, which are of slightly lower resolution ($s_{max} \approx 1/2.5$ nm^{-1}), favor a two-start connectivity model somewhat more strongly. This is seen by examination of the ratio of the best-fitting to the next-best-fitting Q/R values in figure 2. Actin filaments extracted from acto-tropomyosin paracrystals (O'Brien *et al.*, 1983; Morris, 1979) at still higher resolution ($s_{max} \approx 1/2.1$ nm^{-1}) favor the 2′ orientation even more strongly. A close

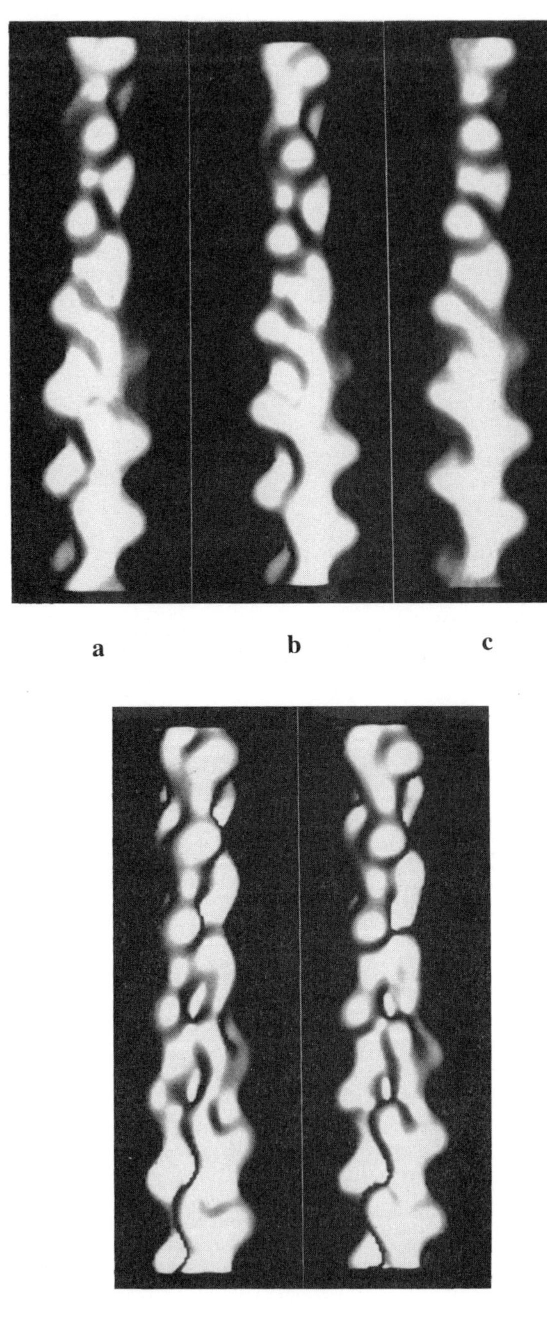

Figure 3. Surfaces of three-dimensional reconstructions of electron microscopy data from F-actin and best-fitting models. a) Model of 2′ fit (like Holmes *et al.*) to Milligan *et al.* data. b) Milligan *et al.* cryo-microscopy data. The resolving power here is near 1/2.9 nm^{-1}. c) Model of 1′ fit (like Schutt *et al.*) to Milligan *et al.* data. d) Morris acto-tropomyosin data with tropomyosin removed. The resolving power of these negatively-stained filaments is near 1/2.2 nm^{-1}. e) Model of best fit (2′) to Morris data.

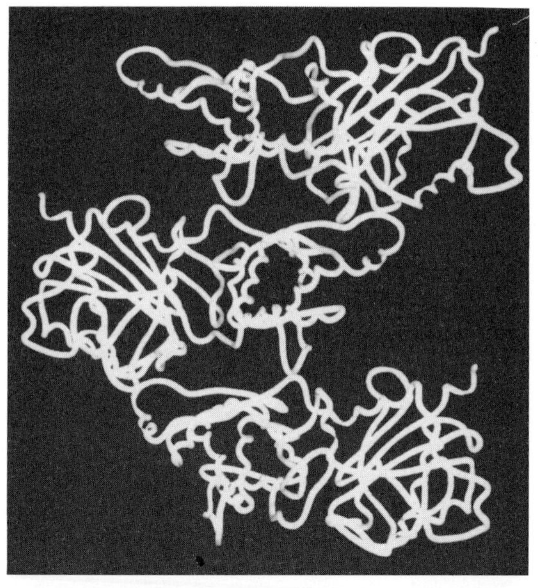

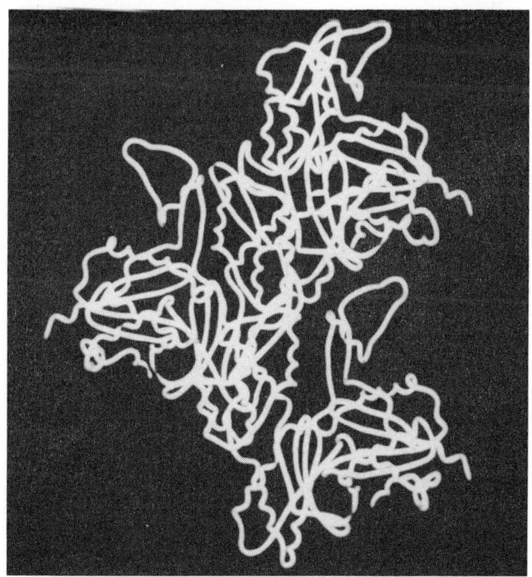

Figure 4. Rubber hose representations of three protomers of F-actin in the 1′ (top) and 2′ (bottom) orientations. The azimuthal orientations are similar to those of the lowest three (full) protomers in figure 3.

correspondence between high-resolution features of the data and its best fit are shown in figures 3d and 3e.

In order to study the effect of resolution on the results obtained by our fitting method, we did a "self-search" on (noiseless) computer-generated layer-line data calculated using the monomer in the Holmes *et al.* orientation. A series of such searches were conducted in the resolution range 1.5 - 3.5 nm. From these searches we were able to recover the same minima to be *expected* that the 1′ orientation would give a good fit at low resolution. Examination of three-dimensional surface views of the data (Fig. 3b) and 1′ (Fig. 3c) and 2′ (Fig. 3a) model filaments reveals why such radically different monomer orientations fit the data almost equally well. At the resolving power of the Milligan *et al.* data, these two models are remarkably similar. On increasing the resolution to near 2.1 nm, the model of the 2′ structure (Fig 3e) shows a distinct cleft between subdomain 4 of one subunit and subdomain 1 of the next subunit across the filament axis. In contrast, the 1′ model (not displayed) continues to show the corresponding regions joined by a prominent ridge of density.

In all of the data sets analyzed using the R-factor, orientations classified as 2′ were favored. The Euler angles which define these orientations are quite similar but are neither identical to those specified by Holmes *et al.* nor to each other. Examination of these different orientations using the molecular graphics program MIDAS (Ferrin *et al.*, 1988) showed this corresponded to a 6°-12° rotation of the actin subunit about an axis lying roughly parallel to the helix axis, so as to move subdomains 1 and 2 closer to the helix axis, and subdomains 3 and 4 further away. In the case of the data from negatively-stained paracrystals, there is an additional component involving a small rotation about the short axis of the actin subunit so that subdomain 2 is moved in towards the helix axis and subdomain 4 is moved out. Although these angular differences seem large, the r.m.s. deviation of atomic positions between structures derived from electron microscopy and those derived from X-ray diffraction was less than 0.35 nm (R_c = 2.35 nm).

The Position of Tropomyosin

The close correspondence of the derived orientations between data sets suggests to us that it might be feasible to compare structures of actin filaments decorated with muscle regulatory proteins in different states. It has been believed for some time that tropomyosin moves circumferentially and perhaps radially on F-actin in response to Ca^{2+} ions binding to troponin (Huxley, 1972; Parry & Squire, 1973). This is thought to allow actin and myosin to interact and thereby generate force. However, proof of this movement of tropomyosin has been difficult to obtain. Neither unstained nor negatively-stained electron microscopies have generated images which clearly and unambiguously show the position of tropomyosin in relaxed thin filaments. However, tropomyosin is clearly resolved in single-layered acto-tropomyosin paracrystals (Morris, 1979; O'Brien *et al.*, 1983).

Tropomyosin, in the absence of troponin, inhibits the ATPase of acto-S1, at least at low S1 concentrations. Ca^{2+} bound to thin filaments activates the ATPase of S1 to levels comparable to that achieved with actin alone under these conditions. Thus, we reasoned that a comparison of these two structures might demonstrate this movement of tropomyosin on F-actin.

In Figure 5 we have aligned the Ca^{2+}-thin-filament (unstained) data of Milligan *et al.* (1990) with the acto-Tm filament from single-layered paracrystals (Morris, 1979; O'Brien, 1983). To achieve this alignment, the resolution of acto-Tm data was lowered to that of the Ca^{2+} thin filaments and longitudinal translations and azimuthal rotations were systematically made to minimize the phase residual (Q) between the two data sets. Since the mass of tropomyosin in the thin filament is about 1/5 of that of actin, the alignment is largely deter-

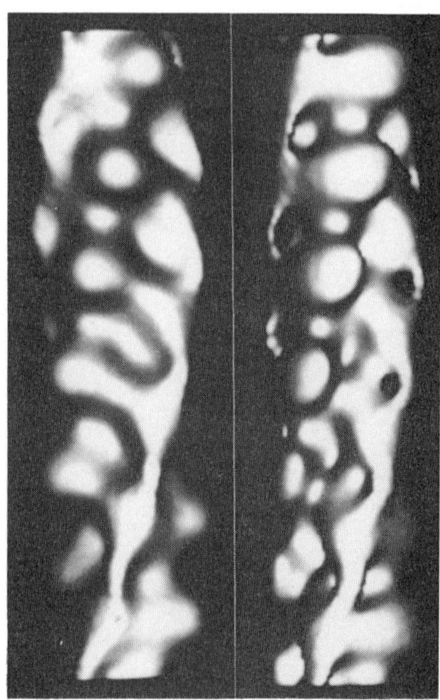

Figure 5. Surface views of three-dimensional reconstructions of Milligan's reconstituted thin filaments with Ca^{2+} present (left) and Morris' acto-tropomyosin filaments. The two have been brought into alignment by a rotation and translation to minimize the phase residuals.

mined by actin. As can be seen, the actin and tropomyosin positions are substantially the same. Close examination of sections from the aligned reconstructions shows that there is a slight circumferential shift of tropomyosin mass away from actin subdomains 1 and 2 and toward subdomains 3 and 4 in the Ca^{2+} thin filaments. We estimate that the change is no more than $10°$. The tropomyosin radii in both images is near 3.8 nm, which agrees with the value of 3.97 ± 1.5 nm obtained by neutron scattering studies of the tropomyosin position in acto-Tm (Bivin *et al.*, 1991). Neutron scattering work (Mendelson *et al.*, 1991) on reconstituted thin filaments, in which only tropomyosin is visible to neutrons, shows that there is little radial change of tropomyosin in response to Ca^{2+}. It is not altogether clear to what extent actin-tropomyosin can be regarded as an analog of the low Ca^{2+} state of the thin filament. However, the inhibition obtained by tropomyosin alone seems to be associated with a position for the tropomyosin which shows a rather small movement from its position in the high Ca^{2+} thin filament. Relatively slight movements of tropomyosin associated with regulation would appear to be in concert with recent views of regulation wherein Ca^{2+} regulates transitions between actin-bound myosin intermediates rather than regulating the binding of the myosin head to actin (Lehrer and Morris, 1982; Geeves and Halsall, 1987).

SUMMARY

The convergence of structures, all determined by independent global searches and subsequent refinement on different electron microscopy data sets, with the X-ray fiber diffraction results strongly suggests that we now have the approximately correct structure for F-actin. This consensus structure will now provide a reliable, well-defined platform upon which to study the structure and function of proteins bound to actin. Among these are capping proteins, such as severin and gelsolin, contractile proteins, such as myosin and its subfragments, and proteins involved in regulation, such as troponin and tropomyosin. A comparison of thin filaments + Ca^{2+} and acto-tropomyosin revealed a slight difference in the

ACKNOWLEDGMENTS

We thank Drs. A. Bremer and R. Milligan for furnishing their electron microscopy data. R.M. is grateful to Dr. E. J. O'Brien for his hospitality during a sabbatical leave at Kings College London and thank him and Prof. K. Holmes and his team at the Max Planck Instut für medizinische Forschung (Heidelberg) for collaboration in an earlier stage of this work. This work was supported by NIH grant R01-AR39710 and NSF grant DMB-8716091 to R.M. We also acknowledge the use of the UCSF Computer Graphics Laboratory, which is supported by NIH, National Center for Research Resources, RR-01081.

REFERENCES

Bivin, D.B., Stone, D.B., Schneider, D.K., and Mendelson, R.A, 1991, Cross-helix separation of tropomyosin as determined by neutron scattering, *Biophys. J.* 59:880.

Bremer, A., Millonig, R.C., Sütterlin, R., Engel, A., Pollard, T.D., and Aebi, U., 1991, The structural basis for the intrinsic disorder of the actin filament: The "lateral slipping" model, *J. Cell. Biol.* 115:689.

DeRosier, D.J. and Moore, P.B., 1970, Reconstruction of three dimensional images from electron micrographs of structures with helical symmetry, *J. Mol. Biol.* 52:355.

Egelman, E.H. and DeRosier, D., 1983, A model for F-actin derived from image analysis of isolated filaments (appendix), *J. Mol. Biol.* 166:623.

Egelman, E.H., 1992, Two key questions raised by an atomic model for F-actin, *Curr. Opin. Struct. Biol.* 2:286.

Elzinga, M. and Phelan, J.J., 1984, F-actin is intermolecularly cross-linked by N,N′-p-phenylene-dimaleimide through lysine-191 and cysteine-374, *Proc. Natl. Acad. Sci. USA* 81:6599.

Ferrin, T.E., Huang, C C., Jarvis, L.E., and Langridge, R., 1988, The Midas display system, *J. Mol. Graphics* 6:13.

Geeves, M.A. and Halsall, D.J., 1987, Two step ligand binding and cooperativity, *Biophys. J.* 52: 215.

Hanson, J., 1967, Axial period of actin filaments, *Nature (Lond.)* 213:353.

Hartt, J. E. and Mendelson, R.A., 1980, X-Ray scattering of F-actin and myosin subfragment 1 complex, *Fed. Proc.* 39:1728.

Hegyi, G., Michel, H., Shabanowitz, J., Hunt, D.F., Chatterjie, N., Healy-Louie, G., and Elzinga, M., 1992, Gln-41 is intermolecularly cross-linked to Lys-113 in F-actin by n-(4-azidobenzoyl)-putrescine, *Prot. Sci.* 1:132.

Holmes, K.C., Popp, D., Gebhard, W. and Kabsch, W. 1990, Atomic model of the actin filament, *Nature (Lond.)* 347:44.

Huxley, H.E., 1972, Structural changes in the actin- and myosin-containing filaments during contraction, *Cold Spring Harbor Symp. Quant. Biol.* 37:361.

Kabsch,W., Mannherz, H.G., Suck, D., Oai, E., and Holmes, K.H., 1990, Atomic structure of actin:DNase I complex, *Nature (Lond.)* 347:37.

Kasprzak, A.A., Takashi, R., and Morales, M.F., 1988, Orientation of actin monomer in the F-actin filament: radial coordinate of glutamine-41 and effect of myosin subfragment 1 binding on the monomer orientation, *Biochem.* 27:4512.

Lehrer, S.S. and Morris, E.P., 1984, Dual effects of tropomyosin and tropomyosin-troponin on acto-myosin subfragment 1 ATPase, *J. Biol. Chem.* 257:8073.

Matsudaira, P., Bordas, J., and Koch, M.H., 1987, Synchrotron x-ray diffraction studies of actin structure during polymerization, *Proc. Natl. Acad. Sci. USA* 84:3151.

Mendelson, R.A., Gebhard, W., Holmes, K., Kabsch, W., Suck, D., Couch, J., Morris, E. and O'Brien, E., 1984, Fitting the actin monomer into the actin helix: a combination of X-ray diffraction and electron microscope data, *Biophys. J.* 45:391a.

Mendelson, R.A., Stone, D.B., Timmins, P.A., and Johnson, E.R., 1991, Results of preliminary studies of tropomyosin radial movement using neutron scattering, *Biophys. J.* 59:219a.

Milligan, R.A. and Flicker, P.F., 1987, Structural relationships of actin, myosin, and tropomyosin revealed by cryo-electron microscopy, *J. Cell Biol.* 105:29.

Milligan, R.A., Whittaker, M. and Safer, D., 1990, Molecular structure of F-actin and location of surface binding sites, *Nature (Lond.)* 348:217.

Morris, E.P., 1979, The effects of troponin components on the structure and activity of thin filaments, *Ph. D. thesis,* University of London.

O'Brien, E.J., Couch, J., Johnson, G.R.P., and Morris, E.P., 1983, Structure of actin and the thin filament, *in*: "Actin: Its Structure and Function in Muscle and Non-muscle Cells" C. Dos Remedios and J. Barden, editors, Academic Press, Sydney, Australia.

Parry, D.A.D. and Squire, J.M., 1973, Structural role of tropomyosin in muscle regulation: analysis of the X-ray diffraction patterns from relaxed and contracting muscles, *J. Mol. Biol.* 75:33.

Popp, D. and Holmes, K.C., 1992, X-ray diffraction studies on oriented gels of vertebrate smooth muscle thin filaments, *J. Mol. Biol.* 224:65.

Schutt, C.E., Lindberg, U., Myslik, J., and Strauss, N., 1989, Molecular packing in profilin:actin crystals and its implications, *J. Mol. Biol.* 209:735.

Schutt, C.E. and Lindberg, U., 1992, Actin as the generator of tension during muscle contraction, *Proc. Natl. Acad. Sci. USA* 89:319.

Suck, D., Kabsch, W., and Mannherz, H.G., 1981, Three-dimensional structure of the complex of skeletal muscle actin and bovine pancreatic DNase I at 6 Å resolution, *Proc. Natl. Acad. Sci. U.S.A.* 78:4319.

Taylor, D., Reidler, L., Spudich, J.A., and Stryer, L., 1981, Detection of actin assembly by fluorescence energy transfer, *J. Cell Biol.* 89:362.

Trinick, J., Cooper, J., Seymour, J., and Egelman, E.H., 1986, Cryo-electron microscopy and three-dimensional reconstruction of actin filaments, *J. Microsc.* 141:349.

EVIDENCE FOR AN F-ACTIN LIKE CONFORMATION IN THE ACTIN:DNASE I COMPLEX

B.D. Hambly[1], P. Kießling, and C.G. dos Remedios

Muscle Research Unit
Department of Anatomy and
[1]Department of Pathology
The University of Sydney NSW 2006 Australia.

ABSTRACT

We demonstrate that a ribose modified analogue of ATP, TNP-ATP, can exchange with a resident nucleotide in F-actin, but fails to bind to G-actin. TNP-ATP is also able to bind to actin in the actin:DNase I complex, suggesting that the nucleotide binding site in the actin:DNase I complex adopts a conformation similar to that found in F-actin. This result is consistent with the hypothesis that the two major domains of actin on either side of the cleft are able to "flex" or move relative to each other in G-actin, but that this flexing motion is limited as a consequence of either polymerisation or DNase I binding. F-actin, in which ~80% of the bound nucleotide is TNP-ADP, appears to be functionally similar to native ADP-F-actin. It can superprecipitate with myosin and, following regulation with troponin-tropomyosin, exhibits a Ca^{2+}-sensitivity during superprecipitation. Sonication induced nucleotide exchange in regulated F-actin was not sensitive to the presence of Ca^{2+} which argues against a significant conformational change in the vicinity of the nucleotide binding site during Ca^{2+}-sensitive thin filament regulation.

INTRODUCTION

The crystal structure of the actin monomer, complexed to DNase I, has recently been determined (Kabsch et al., 1990) and an F-actin structure has been proposed (Holmes et al., 1990). Each monomer consists of two domains, each containing two subdomains. The so-called smaller domain lies at maximal radius in the F-actin structure. A nucleotide/metal complex binds to actin within the cleft between the two domains

(ATP/Ca^{2+} in G-actin and ADP/Ca^{2+} in F-actin). An important question that arises is whether the crystal structure of actin complexed to DNase I more closely resembles the G- or the F-actin structure. Actin in the DNase I complex is thought to be more an analogue of F-actin, since the actin:DNase I interface "mimics" or coincides with one of the predicted actin-actin interactions. Furthermore, the crystal structure of actin is a reasonable fit to the X-ray diffraction pattern derived from oriented gels of F-actin (Holmes et al., 1990). This fitting procedure involved a 6Å shift in the location of sub-domain 2 but no changes were assumed for sub-domains 1, 3 or 4 (Mannherz, 1992). To test the hypothesis that actin in the co-crystal has an F-actin like conformation we have investigated the binding of an analogue of ATP, TNP-ATP, to the actin DNase I complex.

The interaction between F-actin and myosin is required for force generation in a wide variety of tissues. In striated muscle, this interaction is regulated by Ca^{2+} binding to the troponin/tropomyosin (TN-TM) regulatory protein complex of the thin filament (Cooke, 1986). The molecular mechanisms whereby the regulatory proteins modulate the acto-myosin interaction are poorly understood (Zot and Potter, 1987), but may involve a conformational change within actin. This change may involve a relative motion of the two major domains of actin. The nucleotide/metal complex interacts with residues in both of these domains which form the borders of the cleft (Kabsch et al., 1990). Thus, we might expect that Ca^{2+} binding to the regulatory proteins may affect the exchangeability of the nucleotide in actin.

Nucleotide can be removed from both G-actin (Kasai et al., 1965) and F-actin (Barany et al., 1966; Fung and Cooke, 1983) under appropriate conditions without denaturation. Increasing the temperature of the F-actin above 30°C (Asai and Tawada, 1966), partial unfolding in urea (Taniguchi, 1976), sonication (Barany and Finkleman, 1962; Asakura et al., 1963) and myosin binding (Moos and Eisenberg, 1970) all increase the rate of nucleotide exchange in F-actin. This increase is thought to be due to a "loosening" or "fragmentation" of the F-actin structure, resulting in the "exposure" of the bound ADP, thus allowing nucleotide exchange and hydrolysis (Taniguchi, 1976; Oosawa, 1983a). Formation of an actin:DNase I complex impairs, but does not prevent, nucleotide exchange (Polzar et al., 1989).

The binding of TM to the actin filament suppresses the exchange of the bound nucleotide and divalent cation (Kasai and Oosawa, 1969), makes the filaments more rigid (Fujime and Ishiwata, 1971; Oosawa, 1983b) and inhibits spontaneous fragmentation of the filaments (Wegner, 1982). Flexibility of the F-actin-TM-TN complex in the presence of Ca^{2+} is similar to that of F-actin-TM, but in the absence of Ca^{2+} there is a further significant decrease in the flexibility of the thin filament (Ishiwata and Fujime, 1972; Oosawa, 1983b). Actin subunit exchange in F-actin-TM-TN is Ca^{2+} dependent, with the exchange being impaired in the absence of Ca^{2+} (Mihashi, 1984). Ca^{2+} binding to F-actin-TM-TN also alters the orientation of the bound nucleotide (Yanagida and Oosawa, 1980). It is therefore possible that this Ca^{2+} sensitive change in the orientation of the nucleotide is associated with a conformational change in actin and with a change in the rate of nucleotide exchange in regulated F-actin.

The affinity of actin for ATP is not affected by modification of the adenine ring of ATP (Barden and dos Remedios, 1984), but it is affected by modification of the ribose moiety, as in TNP-ATP (dos Remedios & Cooke, 1984; dos Remedios et al., 1985; Miki et al., 1986; dos Remedios et al., 1985). We aim to exploit the relatively poor affinity of TNP-ATP for G-actin by using the exchange of this nucleotide as a marker for conformational changes within the actin molecule, both as a result of DNase I binding and during Ca^{2+} regulation of F-actin.

MATERIALS AND METHODS

Reagents and Protein Preparation

TNP ATP was purchased from Molecular Probes Inc. All other chemicals were analytical grade. Actin was prepared from rabbit skeletal muscle acetone powder as described by Barden and dos Remedios (1984). Myosin was prepared using the method of Tonomura et al. (1966). The TN-TM complex was prepared using the method of Hartshorne and Mueller (1969). F-actin was regulated by adding the TN-TM complex to F-actin at a molar ratio of 10:1. The complex was incubated for 60 min. at 4°C and sedimented at 200,000 x g for 100 min. SDS-PAGE gels of the regulated F-actin showed that the TN-TM complex was bound stoichiometrically to and sedimented with F-actin. Protein concentrations were determined using absorption coefficients for actin at 290nm of 0.63mg-1cm-1, for myosin at 280nm of 0.56mg-1cm-1, for the TN-TM complex at 280nm of 0.60mg-1cm-1, of regulated F-actin at 280nm of 0.88mg-1cm-1 and of TNP-ATP at 408nm of 26,400M-1cm-1. Relative molecular masses of 42.3kDa for actin, 500kDa for myosin, 136kDa for the TN-TM complex were used.

TNP-ATP Binding to G- and F-actin

Free nucleotide was removed from a solution containing ~50μM G-actin dissolved in 0.2mM ATP, 2mM DTT, 2mM NaN3 and 2mM Tris pH 7.0, by the addition of 1/10 volume of Dowex AG 1-X4 (BioRad), which was then removed by sedimentation. TNP-ATP was added to G-actin at increasing molar excesses of TNP-ATP over G-actin ranging between 0 and 50-fold. The samples were incubated at 22°C for 150 min. and were then allowed to polymerise for 15 min in 50mM KCl, 2mM MgCl2 and 0.02M sodium phosphate buffer, pH 7.0. Dowex-1 treatment, repeated once, was used to remove free TNP-ATP. A second set of samples was treated in the same way, except that G-actin was not polymerised. Incorporation of TNP-ATP into this G-actin was then determined spectrophotometrically.

TNP-ATP Incorporation into F-actin by Sonication

Free nucleotide was removed using Dowex-1 from a sample of ~50μM F-actin or regulated F-actin dissolved in 0.1M KCl, 2mM MgCl2 and 20mM HEPES pH 7.0. A 50-molar excess of TNP-ATP over actin was added to 2ml of this solution, which was then sonicated for up to 15 min in an ice bath, using an Ultrasonics W-70 sonicator operated at maximum power, for bursts of 5 s duration with 5 s rests. 0.175ml aliquots were taken at various time intervals during sonication and sedimented in a Beckman Airfuge operated at room temperature at 200kPa for 15 min and the pellet was washed in buffer. The resuspended pellet was sedimented a second time to minimise free TNP-ATP in the solution. The incorporation of TNP-ADP into the resuspended F-actin samples was then determined spectrophotometrically.

TNP-ATP Exchange into the G-actin:DNase I Complex

Free nucleotide was removed by Dowex-1 treatment from ~50μM G-actin or G-actin-DNase I complex dissolved in 0.2mM ATP, 2mM DTT, 2mM NaN3 and 2mM Tris, pH 7.0. A 20-molar excess of TNP-ATP over actin was added to each of these solutions and incubated at 22°C for 150 min. At the completion of the incubation, a 2-molar excess of DNase I was added to the sample. Dowex-1 treatment, repeated once, removed free TNP-ATP and the incorporation of TNP-nucleotide was determined spectrophoto-metrically.

Superprecipitation Experiments

A sample of ~80% TNP-ADP labelled F-actin (50µM) was prepared by sonication incorporation. Half the sample was regulated with the TN-TM complex. Two similar samples of unlabelled F-actin and regulated F-actin (~50µM) were prepared as controls. Skeletal muscle myosin was added to each of the actin samples in the ratio 1:2.5 w/w of actin:myosin, diluted to 0.15mg/ml total protein in 25mM KCl, 1mM $MgCl_2$, 0.2mM DTT, 0.2mM EGTA and 20mM HEPES pH 7.0. The rate ($T_{1/2}$) and extent (ΔOD) of superprecipitation was determined from the change in the optical density in a stirred 2.0cm path cuvette at 550nm and 25°C. The mixing time of the 10ml sample was 1.8 s. Superprecipitation was initiated in the stirred suspension of contractile proteins by the addition of ATP to 0.1mM. Superprecipitation of regulated actomyosin was initiated by the addition of 1mM $CaCl_2$ subsequent to ATP inclusion.

Cross-Linking Experiments

Cross-linking reactions were performed according to Mornet et al. (1981) using 5mM 1-ethyl-3(3-dimethylamonoproyl)carbodiimide as cross-linker in 100mM MES, pH 6.0 at room temperature. Rigor complexes of actin in the actin:DNaseI complex or F-actin and myosin subfragment-1 were pre-formed at an equimolar ratio. The cross-linking reaction was stopped by suspending an aliquot in boiling Laemmli sample-buffer and analyzed by examining the time course of the cross-linking pattern by SDS-PAGE.

RESULTS

TNP-Nucleotide Incorporation into G-actin, F-actin and Actin-DNase I Complex

Fig. 1 illustrates the fractional incorporation of TNP-ATP into the G- and F-actin samples as a function of the molar excess of TNP-ATP added to the samples. 50-molar excess of TNP-ATP resulted in approximately 80% incorporation of TNP-ADP in F-actin. A 20-molar excess of TNP-ATP produced 43% incorporation in F-actin but only 3% in G-actin.

Table 1 shows the percentage incorporation of TNP-ATP into G-actin and G-actin complexed to DNase I using a 20-molar excess of TNP-ATP. TNP-ATP constituted only 4% of the bound nucleotide after exchange in G-actin, whereas the presence of DNase I bound to G-actin during the exchange procedure increased the proportion of bound TNP-ATP to 25%. However, when DNase I was added to G-actin already equilibrated with TNP-ATP, the proportion of TNP-ATP bound to the actin DNase I complex increased to 80%. The incorporation of TNP-ATP into the actin-DNase I complex was fully reversed by the addition of 1mM ATP and subsequent sonication for 60 s (data not shown).

TNP-ADP Incorporation into F-Actin and Regulated F-Actin

Fig. 2 illustrates the percentage of TNP-ADP incorporated into F-actin as a function of sonication time for both F-actin and regulated F-actin. 50% incorporation was achieved in F-actin after approximately 3 min sonication, and approximately 100% incorporation was obtained after 6 min. 50% incorporation into regulated F-actin was only achieved after

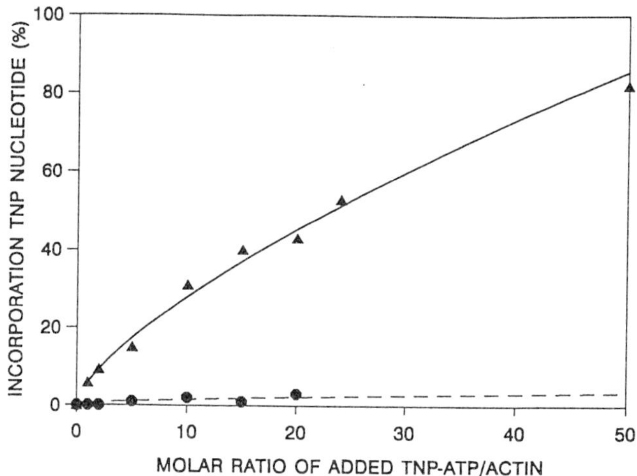

Figure 1. Incorporation of TNP-ATP/ADP into samples of G- and F-actin as a function of the molar excess of TNP-ATP added to the samples. TNP-ATP was added to G-actin samples and, after 150 min, half of each sample was polymerised (▲) while the other half remained in the form of G-actin (●). The percentage incorporation of TNP-ATP/ADP was determined after removal of free nucleotide with Dowex-1. With a 20-molar excess, 43% incorporation was achieved in F-actin compared to only 3% in G-actin.

Table 1. Incorporation of TNP-ATP into Actin.

Form of Actin	TNP-ATP/ADP Incorporation per Actin
G-actin	3%
F-actin	43%
Actin:DNase I + TNP-ATP	30%
G-actin +TNP-ATP+ DNase I	80%

Table 1: Samples of G-actin or actin:DNase I complex were incubated in a 20-molar excess of TNP-ATP for 150 min. A 2-molar excess of DNase I was then added to one of the G-actin samples and samples were polymerised. Free TNP-ATP was removed using Dowex-1 and the percentage incorporation of TNP-ATP/ADP in each of the samples was determined spectrophotometrically. Polymerisation of actin or DNase I binding to actin substantially increases the binding of TNP-ATP to actin.

approximately 8 min, and a further sonication to a total of 15 min did not significantly increase this. Clearly, the binding of the regulatory proteins to F-actin impairs the incorporation by sonication of TNP-ATP into F-actin. The incorporation of TNP-ADP into F-actin was fully reversed by the addition of 1mM ATP and subsequent sonication for 60 s (data not shown).

TNP-ATP was incorporated by sonication into regulated F-actin while the free Ca^{2+} concentration was controlled by EGTA. After a brief Dowex-1 treatment, either EGTA (0.64mM) or $CaCl_2$ (1mM) was added. Fig. 3 shows the percentage incorporation of

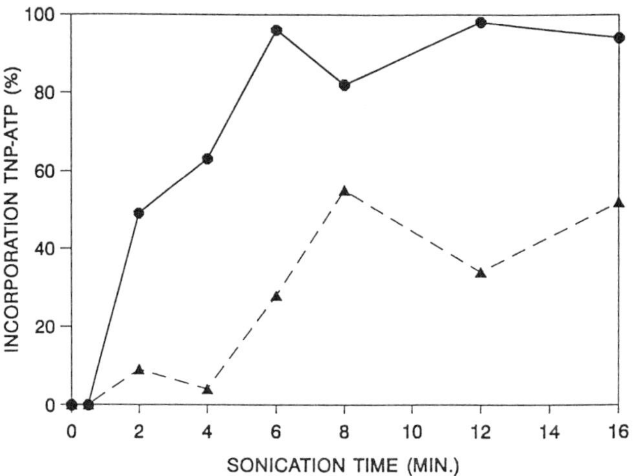

Figure 2. Incorporation of TNP-ADP into F-actin as a function of sonication time for both F-actin (●) and regulated F-actin (▲). 50% incorporation was achieved in F-actin after approximately 3 min sonication, with approximately 100% incorporation being obtained after 6 min. 50% incorporation into regulated F-actin was achieved after approximately 8 min. Further sonication to a total of 15 min did not improve the incorporation significantly, indicating that the binding of the regulatory proteins to F-actin impairs the incorporation by sonication of TNP-ATP in F-actin.

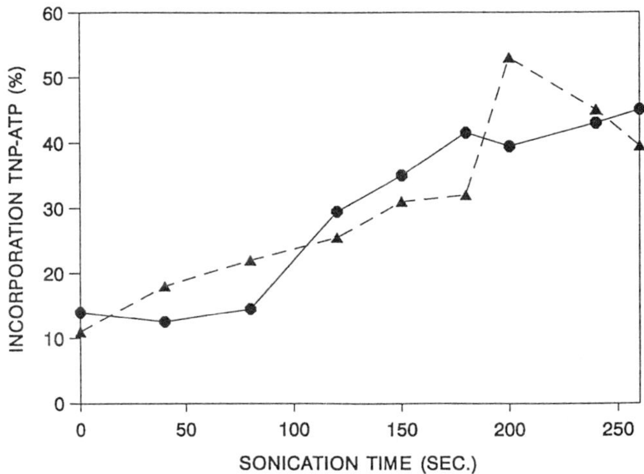

Figure 3. The percentage incorporation of TNP-ADP into regulated F-actin as a function of sonication time in the presence (●) and absence (▲) of free Ca^{2+}. Free Ca^{2+} concentration was controlled by buffering with EGTA. Either EGTA to 0.64mM or $CaCl_2$ to 1mM was added to regulated F-actin prior to the addition of TNP-ATP and sonication. No significant difference was seen between the levels of incorporation between these two samples. Approximately 50% incorporation was achieved after approximately 4 min accumulated sonication time.

TNP-ADP into regulated F-actin in the presence and absence of free Ca^{2+} as a function of sonication time. No difference was detected in the level of incorporation between these two samples under these conditions.

Superprecipitation Experiments

Superprecipitation was used to examine the effect of TNP-ATP binding to both regulated and unregulated F-actin on its ability to bind myosin, stimulate the myosin ATPase and regulate the acto-myosin interaction. F-actin containing approximately 80% TNP-ADP exhibited no significant difference in the extent of acto-myosin superprecipitation compared to unlabelled acto-myosin. However, the rate ($T_{1/2}$) of this reaction decreased approximately 2.5 fold from 4.5 s (unlabelled) to 12 s (TNP-ADP labelled). By contrast, no superprecipitation was detected when ATP was added to regulated actomyosin, regardless of TNP-ADP incorporation. The addition of free Ca^{2+} to the solution was necessary to initiate superprecipitation of TM-TN regulated F-actin. The rate and extent of superprecipitation was similar for the two regulated F-actin samples.

In summary, TNP-ADP labelled F-actin is capable of binding myosin and activating its ATPase activity, although the rate of superprecipitation is reduced. Additionally, the regulatory proteins are capable of binding TNP-ADP labelled F-actin and imparting complete Ca^{2+}-sensitivity to the reconstituted thin filaments.

Cross-linking Experiments

Subfragment-1 incubated with actin:DNaseI can be chemically cross-linked to actin, using EDC as the cross-linker (Fig. 4). A doublet band at 180kDa Mr become visible on SDS-PAGE, after 30 min incubation, indicating the formation of an acto-S1 complex similar to that obtained when F-actin is used instead of actin:DNaseI (Fig. 4).

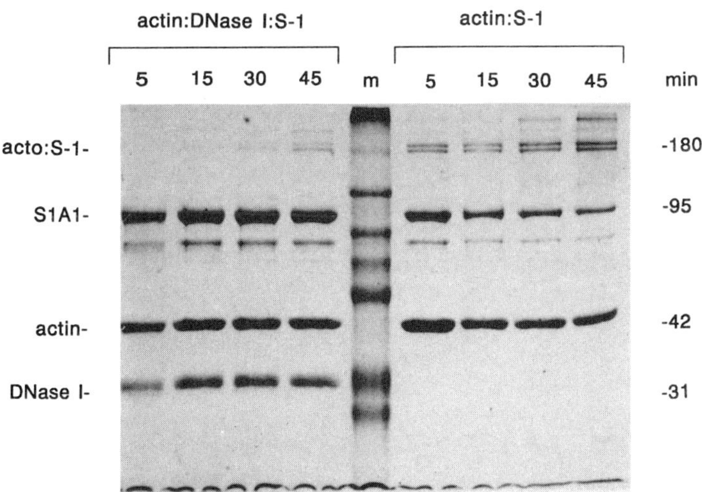

Figure 4. Time course of cross-linking patterns of actin DNase I (lane a-d) and F-actin (lane f-i) to S-1 with EDC (5 mM). Aliquots were suspended in boiling Laemmli sample buffer at the time indicated and subjected to SDS-PAGE. Numbers on the side are molecular masses in kilodalton. Standard molecular marker (m).

DISCUSSION

Our data show that TNP-ATP can be incorporated into the nucleotide binding site of F-actin by both sonication and polymerisation. By contrast, TNP-ATP failed to incorporate into G-actin. Like ATP, TNP-ATP is hydrolysed during the formation of the actin polymer (dos Remedios and Cooke, 1984). TNP-ADP incorporation into F-actin can be fully and readily reversed by sonication in excess ATP, implying that the affinity of TNP-ADP for F-actin is lower than the native nucleotide. These data are consistent with previous findings that modification of the ribose impairs binding of the nucleotide to actin (Martonosi, 1962; Kuroda & Maruyama, 1972). The incorporation of TNP-ADP into the nucleotide site of actin does not appear to functionally change the conformation of the actin molecule. TNP-ADP exchanged actin can polymerise, this polymerised F-actin can superprecipitate, which demonstrates that TNP-ADP F-actin can bind myosin and that the activation of the myosin ATPase can be fully regulated by the regulatory proteins.

Complexing of DNase I to G-actin increased the extent of incorporation of TNP-ATP into actin. In agreement with previous work, we found that DNase I binding to actin does not prevent nucleotide exchange, although others have shown that nucleotide exchange is impaired following DNase I binding (Polzar et al., 1989). Additionally, these authors have shown that the actin:DNase I complex can incorporate either ATP or ADP, but not AMP. The data in Table 1 suggest a model of TNP-ATP exchange into G-actin in which TNP-ATP readily binds to G-actin, as an unstable complex, probably because of a high dissociation rate. Hence, Dowex-1 treatment of G-actin following TNP-ATP exchange removes essentially all TNP-ATP from G-actin. However, if the conformation of actin is altered, either by polymerisation or DNase I binding, then TNP-ATP is not easily removed from actin by a subsequent Dowex-1 treatment. The similar ability of TNP-ADP to remain bound in F-actin and in the actin:DNase I complex supports the hypothesis that the actin nucleotide locus has a similar conformation in these two complexes. Using the actin:DNase I complex as a model system for monomeric actin, F-actin like features, such as nucleotide sensitive S-1-binding and proteolytic protection of an S-1 tryptic cleavage site, could be demonstrated. However, no activation of the myosin S-1 ATPase by actin:DNase I can be detected (Kießling et al., 1992). These results, taken together with the fact that the actin:DNase I complex could contain either ATP, like G-actin or ADP, like F-actin (Polzar et al., 1989), point to a more "chimeric" or intermediate conformation of actin:DNase I, that shares conformational properties of both G- and F-actin.

The actin crystal structure shows that the nucleotide binds to residues in both major domains of actin, deep in the cleft formed by these two domains (Kabsch et al., 1990). The adenine of the nucleotide does not bind to specific amino acids, while the 2'- and 3'-hydroxyl groups of the ribose form hydrogen bonds with Glu 214 and Asp 157 respectively. TNP-ATP is modified by its trinitrophenyl moiety which resonates as a Meisenheimer complex to the 2'- and 3'-hydroxyl positions of the ribose. Since other ribose- modified ATP analogues (e.g. MANT-ATP; unpublished data) do bind to G-actin, albeit with a lower affinity, it may be that the resonating 2' or 3' structure of TNP-ATP is a key element in its impaired binding to the nucleotide site of actin. The phosphate groups of the nucleotide form multiple hydrogen bonds to both domains deep in the cleft and to the metal bound to the high affinity metal site. On the basis of the actin crystal structure, the two domains of actin appear to be rather flexible. This point was also made by another paper in this volume (M. Tirion), who found that the two large domains of actin flex relative to each other. Furthermore, binding of DNase I may result in a distortion of the actin nucleotide site (Kabsch et al., 1990). Binding of DNase I may be associated with a narrowing of the cleft between these two domains, as well as a steric blocking of the ATP cleft. This may also occur during the polymerisation process (Kabsch et al., 1990). TNP-ATP may only form bonds to actin by its phosphate side chain. Given that TNP-ATP fails to form a strong bond with G-actin, it seems that the bonds formed through the phosphate groups of the nucleotide are relatively weak when actin is in its globular monomeric conformation. Flexibility between the two domains in G-actin may also make the bound nucleotide more accessible to the solvent and hence to Dowex-1 removal, thus providing an explanation for the ease with which TNP-ATP is removed from G-actin.

Our TNP-ATP data agree with other studies of nucleotide exchange, namely that regulation of the F-actin filament results in a significant decrease in the exchange of nucleotide into the nucleotide site. This decrease in exchange rate may be ascribed to several factors. Firstly, F-actin may impair exchange by increasing the rigidity and resistance to fragmentation of the filament that results from the binding of TM in the two long-pitch grooves. Similarly, the presence of bound TM may also reflect a stabilisation of flexing between the two domains of actin. Based on previous data, we expected that the Ca^{2+}-sensitive changes observed in regulated thin filament rigidity (Ishiwata and Fujime, 1972), actin subunit exchange (Mihashi, 1984) and orientation of the bound nucleotide (Yanagida and Oosawa, 1980) would be reflected in Ca^{2+}-sensitive changes in nucleotide exchange in regulated F-actin. In fact, we saw no change in nucleotide exchange in response to Ca^{2+} regulation. This raises doubts about significant conformational changes in the vicinity of the nucleotide binding site during Ca^{2+} sensitive regulation of the thin filament.

Acknowledgments

This research was supported by grants from the NH&MRC of Australia.

REFERENCES

Asai, H. & Tawada, K., 1966, Enzymic nature of F-actin at high temperature, *J. Mol. Biol.* 20:403-417.

Asakura, S., Taniguchi, M. & Oosawa, F., 1963, The effect of sonic vibration on exchangeability and reactivity of the bound adenosine diphosphate of F-actin, *Biochim. Biophys. Acta* 74:140-142.

Barany, M. & Finkelman, F., 1962, The lability of F-actin-bound calcium under ultrasonic vibration, *Biochim. Biophys. Acta* 63:98-105.

Barany, M., Tucci, A.F. & Conover, T.E., 1966, Removal of the bound ADP of F-actin. *J. Mol. Biol.* 19:483-502.

Barden, J.A. & dos Remedios, C.G., 1984, The environment of the high affinity cation binding site on actin and the separation between cation and ATP sites as revealed by proton NMR and fluorescence spectroscopy, *J. Biochem.* 96:913-921.

Cooke, R., 1972, A new method for producing myosin subfragment-1, *Biochem. Biophys. Res. Commun.* 49:1021-1028.

Cooke, R., 1986, The mechanism of muscle contraction, *CRC Crit. Rev. Biochem.* 21:53-118.

dos Remedios, C.G. & Cooke, R., 1984, Fluorescence energy transfer between probes on actin and probes on myosin, *Biochim. Biophys. Acta* 788:193-205.

dos Remedios, C.G., Hambly, B.D. & Barden, J.A., 1985, On the nucleotide binding site of actin, *in:* "Cell Motility: Mechanism and Regulation", p. 31-38, H. Ishikawa, S. Hatano & H. Sato eds., University of Tokyo Press, Tokyo.

Fujime, S. & Ishiwata, S., 1971, Dynamic study of F-actin by quasi-electric scattering of laser light, *J. Mol. Biol.* 62:251-265.

Fung, B. & Cooke, R., 1983, Studies on the role of the actin nucleotide in contraction and polymerisation, *in:* "Actin Structure and Function in Muscle and Non-Muscle Cells," C.G. dos Remedios and J.A. Barden, eds., Academic Press, Sydney.

Hartshorne, D.J. & Mueller, H., 1969, The preparation of tropomyosin and troponin from natural actomyosin, *Biochim. Biophys. Acta* 175:301-319.

Holmes, K.C., Popp, D., Gebhard, W. & Kabsch, W., 1990, Atomic model of the actin filament, *Nature* 347:44-49.

Ishiwata, S. & Fujime, S., 1972, Effect of calcium ions on the flexibility of reconstituted thin filaments of muscle studied by quasi-electric scattering of laser light, *J. Mol. Biol.* 68:511-522.

Kabsch, W., Mannherz, H.G., Suck, D., Pai, E.F. & Holmes, K.C., 1990, Atomic structure of the actin:DNase I complex, *Nature* 347:37-44.

Kasai, M., Nakano, E. & Oosawa, F., 1965, Polymerisation of actin free from nucleotides and divalent cations, *Biochim. Biophys. Acta* 94:494-503.

Kasai, M. & Oosawa, F., 1969, Behaviour of divalent cations and nucleotides bound to F-actin, *Biochim. Biophys. Acta* 172:300-310.

Kießling, P., Schick, B., Polzar, B. & Mannherz, H.G., 1992, The interaction of myosin subfragment-1 with monomeric actin, *J. Muscle Res. Cell Motil.* 13: 263.

Kuroda, M. & Maruyama, K.,1972, Polymorphism of F-actin. II. ATPase activity at acid pH, *J. Biochem.* 71:39-45.

Martonosi, A., 1962, Deoxy ATP (or GTP) do not bind to G-actin, *Biochim. Biophys. Acta* 57:163.

Mannherz, H.G., 1992, Crystallization of actin in complex with actin-binding proteins, *J. Biol. Chem.* 267:11661-11664.

Mihashi, K., 1984, Ca^{2+}-dependent regulation of the subunit exchange of F-actin under the influence of tropomyosin and troponin. *J. Biochem.* 96:273-276.

Miki, M., Hambly, B.D. & dos Remedios, C.G., 1986, Fluorescence energy transfer between nucleotide binding sites in an F-actin filament, *Biochim. Biophys. Acta* 871:137-141.

Moos, C. & Eisenberg, E., 1970, Effect of myosin on actin-bound nucleotide exchange in the presence and absence of ATP, *Biochim. Biophys. Acta* 223:221-229.

Oosawa, F., 1983a, Macromolecular assembly of actin, *in:* "Muscle and Nonmuscle Motility." Vol. 1, A. Stracher, ed., Academic Press, New York.

Oosawa, F., 1983b, Physicochemical properties of single filaments of F-actin, *in:* "Actin Structure and Function in Muscle and Non-muscle Cells", p. 69-80, C.G. dos Remedios & J.A. Barden, eds., Academic Press, Sydney.

Polzar, B., Nowak, E., Goody, R.S. & Mannherz, H.G., 1989, The complex of actin and deoxyribonuclease I as a model system to study the interactions of nucleotides, cations and cytochalasin D with monomeric actin, *Eur. J. Biochem.* 182:267-275.

Taniguchi, M., 1976, Diphasic transformation of F-actin. Effects of urea and magnesium chloride on F-actin, *Biochim. Biophys. Acta* 427:126-140.

Tonomura, Y., Appel, P. & Morales, M.F., 1966, On the molecular weight of myosin II, *Biochemistry* 5:515-521.

Wegner, A., 1982, Kinetic analysis of actin assembly suggests that tropomyosin inhibits spontaneous fragmentation of actin filaments, *J. Mol. Biol.* 161:217-227

Yanagida, T. & Oosawa, F., 1980, Conformational changes of F-actin-ε-ADP in thin filaments in myosin-free muscle fibres induced by calcium, *J. Mol. Biol.* 140:313-320.

Zot, A.S. & Potter, J.D., 1987, Structural aspects of troponin-tropomyosin regulation of skeletal muscle contraction, *Ann. Rev. Biophys. Chem.* 16:535-559.

ACTIN-BOUND NUCLEOTIDE/DIVALENT CATION INTERACTIONS

Lewis C. Gershman, Lynn A. Selden, Henry J. Kinosian, and
James E. Estes

Research and Medical Services
Stratton VA Medical Center
Albany, New York 12208
            and
Departments of Medicine and Physiology and Cell Biology
Albany Medical College
Albany, New York 12208

INTRODUCTION

The protein actin is a major constituent of the cytoskeleton of virtually all
eukaryotic cells. The relatively stable actin filament structure of muscle cells is
essential to muscle contraction; the motile events of non-muscle cells involve active
reorganization of the actin filament network. Actin has binding sites for one tightly
bound divalent cation and one adenosine nucleotide per molecule. These ligands
are important to the stability of the actin molecule and actin filament, and to the
ATP hydrolysis that is involved in actin polymerization in a way which is still incompletely understood.

Until the recent publication (Kabsch et al., 1990) of the atomic structure of
the actin:DNase I complex made the situation significantly clearer, the nature and
interrelationship of the high affinity binding of divalent cation and adenosine
nucleotide to actin were poorly understood. Furthermore, despite early indications
of differences in actin conformation and polymerization characteristics depending
upon the type of divalent cation and adenosine nucleotide bound (see Estes et al.,
1992, for review), in many studies on actin over many years little attention was paid
to the nature and influence of these ligands. Indeed, in reviewing the actin literature, it is frequently difficult to gauge whether experiments actually reflect actin
containing Ca^{++} (Ca-actin) or Mg^{++} (Mg-actin) at the high affinity site or whether
some combination of the two was present.

The high affinity binding of divalent cation to actin has recently been
extensively reviewed (Estes et al., 1992). More recently, the dependence of nucleotide binding on the divalent cation has been clarified, and a pseudo-first order

model for the competitive exchange of nucleotide on actin has been established. (Kinosian et al., 1993). The present review draws heavily from the ideas and data published in these two manuscripts; however, it also presents some theoretical insights and new data bearing on the incorporation of Ca-actin during attempts to prepare Mg-actin polymers. The paper is primarily directed toward identifying the problems and pitfalls which arise when attempting to manipulate the high affinity divalent cation and nucleotide on actin.

It should be noted that the following discussion on divalent cation and nucleotide binding to actin refers to binding by actin monomer. The tightly bound divalent cation and nucleotide in polymer actin have long been considered unexchangeable (Kasai and Oosawa, 1969; Estes and Moos, 1969); indeed, polymerization has been used to stabilize divalent cation binding for measurement (e.g., Gershman et al., 1986). Recent studies (Newman et al., 1993) suggest that nucleotide exchange on polymer is at least an order of magnitude slower than on monomer actin. It is possible that the nucleotide within polymer is unexchangeable, and that any exchange observed is due to monomer cycling at the polymer ends.

TIGHTLY BOUND DIVALENT CATION

Background

Early on, the tightly-bound divalent cation in preparations of actin was identified as Ca^{++} (Maruyama and Gergely, 1961). Furthermore, it was recognized that the tightly-bound Ca^{++} was freely exchangeable with other metal ions (for example, see Kasai and Oosawa, 1969). From studies of EDTA-washed myofibrils (Weber et al., 1969), and from electron probe analysis of frog skeletal muscle (Kitizawa et al., 1982), it appears that the tightly-bound divalent cation *in vivo* is most likely Mg^{++}. However, Ca-actin continues to be the standard actin preparation and experiments are still performed on actin whose tightly-bound divalent cation is poorly defined.

Divalent Cation Binding Affinity

The somewhat convoluted history leading to our present understanding of the high affinity binding of divalent cations to actin has recently been reviewed by Estes et al. (1992). In brief, based on equilibrium dialysis and ultracentrifugal studies, the dissociation constant for high affinity binding of Ca^{++} or Mg^{++} to actin was initially estimated to be in the 1.0-10.0 µM range (Martinosi et al., 1964). A value in this range was supported (Frieden et al., 1980; Frieden, 1982) by results based on an "isomerization model" for exchange of divalent cation on actin which has subsequently been disproved (Estes et al., 1992). Then Konno and Morales (1985) reported that two thiol groups in actin, normally unreactive, became exposed at very low Ca^{++} concentrations, suggesting that the Ca^{++}-binding affinity of actin might be much stronger than previously thought. That this is the case was demonstrated shortly thereafter (Gershman et al., 1986), and subsequently confirmed by others (see review by Estes et al., 1992).

An experiment which directly shows the high affinity of actin for Ca^{++} and also gives an indication of the slowness of divalent cation exchange kinetics is illustrated in Figure 1 (reprinted from Gershman et al., 1986; Estes et al., 1992). In this experiment, the fluorescent Ca^{++} chelator Quin 2 is used to induce and monitor the dissociation of tightly-bound Ca^{++} from actin. The slow increase in fluorescence intensity up to the first plateau is due to the release of Ca^{++} from the high affinity site. The fluorescence intensity then increases to the second plateau as the remaining tightly-bound Ca^{++} is displaced by added Mg^{++}. It is evident that at the first plateau approximately one half the tightly-bound Ca^{++} has been dissociated; the equilibrium dissociation constant (K_{Ca}) for tightly-bound Ca^{++} is thus equal to the free Ca^{++} concentration present at the first plateau, which was calculated to be ~ 2 nM.

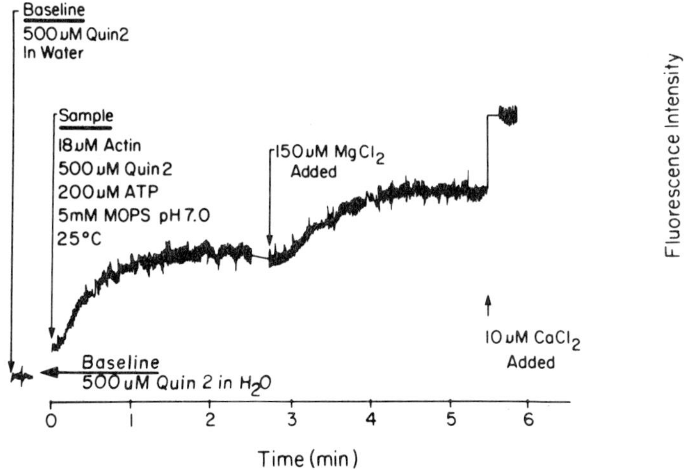

Figure 1. Time course of Ca^{++} dissociation from the high affinity site on actin. The fluorescence intensity of Quin 2 versus time is shown. Sample fluorescence intensity increases to the first plateau as Ca^{++} dissociates from actin and binds to Quin 2. After addition of $MgCl_2$, the fluorescence intensity increases to the second plateau as the remaining actin-bound Ca^{++} is displaced by Mg^{++}. Final "step" shows increase in fluorescence with addition of 10 µM Ca^{++} calibration. Figure previously published in Gershman et al. (1986) and Estes et al. (1992), and reappears here with permission.

The approximate value for K_{Ca} obtained from the experiment of Figure 1 is in excellent agreement with the results of extensive studies which put the consensus value for K_{Ca} at 2 nM at pH 7.0 (Estes et al., 1992). K_{Ca} is approximately 4 nM at pH 7.5 and 8 nM at pH 8.0 (Estes et al., 1987). The equilibrium constant (K_{Mg}) for dissociation of tightly-bound Mg^{++} from actin has been estimated from competitive binding experiments (Gershman et al., 1986; Estes et al., 1987); K_{Mg} is about 9 nM at pH 7.0, and increases to 17 nM at pH 7.5 and 28 nM at pH 8.0. The absolute values for these dissociation constants are ultimately referenced to the dissociation constant for Ca-Quin 2. The estimates for K_{Ca} and K_{Mg} given above were based on a K_D for Ca-Quin 2 of 60 nM at pH 7.0 and 0.1 ionic strength (Tsien et al., 1982). More recently we measured the K_D for Ca-Quin to be $\approx$ 37 nM at the low ionic strength conditions under which most of our binding studies were originally done (Gershman et al., 1991). Thus K_{Ca} should be revised downwards to $\approx$ 1 nM, and K_{Mg} to $\approx$ 5 nM at pH 7.0. Actin binds divalent cation with higher affinity than all other known Ca^{++} or Mg^{++} binding proteins.

Binding Kinetics

For binding of this high affinity, even assuming diffusion limited association kinetics, one would expect the dissociation kinetics to be slow. Figure 1 suggests that this is indeed the case. The half-time for the fluorescence increase to the second plateau[1] is about 40 sec; from this, one may estimate the rate constant for dissociation of tightly bound Ca^{++} from actin to be $k_{-Ca} \sim 0.017$ sec. This value is in excellent agreement with the consensus value at pH 7.0, $k_{-Ca} \approx 0.014$ sec^{-1} (Estes et al., 1992). In parallel with K_{Ca} and K_{Mg}, k_{-Ca} and k_{-Mg} increase at higher pH (Estes et al., 1987), and with higher ionic strength or secondary salt binding (Selden et al., 1989).

The binding kinetics for Mg^{++} are somewhat different. Although Mg^{++} binds to actin with only 1/5 the affinity of Ca^{++}, the dissociation rate constant for Mg^{++} (k_{-Mg}) is not larger than k_{-Ca} but is only about 1/10 the value of k_{-Ca} (Estes et al., 1987). This implies that k_{Mg}, the association rate constant for Mg^{++}, must be smaller than k_{Ca}, the association rate constant for Ca^{++}, by a factor of 50 or so. Determinations of these rate constants (Gershman et al., 1991) showed that this is indeed the case. k_{Ca} was measured to be $\sim 2 \times 10^{7}$ $M^{-1}sec^{-1}$ which, allowing for the geometry of the binding site, probably reflects diffusion limited association. k_{Mg} was measured to be $\sim 2.5 \times 10^{5}$ $M^{-1}sec^{-1}$, much too small for diffusion limited association. The low values for k_{Mg} and k_{-Mg} reflect the slow kinetics of the Mg^{++} aquo-ion. The inner hydration shell of the Mg^{++} aquo-ion is tightly bound with a rate constant for dehydration on the order of 1.5×10^{5} sec^{-1} (Diebler et al., 1969). This slow dehydration reaction limits the rate for association reactions of Mg^{++} in its aquo-ion form (see Gershman et al, 1991, for more details). Thus the kinetics of the binding of Ca^{++} to the high affinity site on actin are diffusion limited, but the kinetics of Mg^{++} binding are limited by the characteristics of the Mg^{++} aquo-ion.

The measured kinetics for binding of Ca^{++} and Mg^{++} to actin are consistent with the values for K_{Ca} and K_{Mg} reviewed in the previous section. Calculating from $K_{Ca} = k_{-Ca}/k_{Ca}$, we find $K_{Ca} = (0.14$ $sec^{-1})/(2 \times 10^{7}$ $M^{-1}sec^{-1}) = 0.7$ nM, in good agreement with the value of 1 nM from the more direct binding estimates. Similarly, $K_{Mg} = k_{-Mg}/k_{Mg} = (.0012$ $sec_{-1})/(2.5 \times 10^{-5}$ $M^{-1}sec^{-1}) = 5$ nM, in excellent agreement with the estimate of 5 nM from equilibrium binding measurements.

Divalent Cation Exchange On Actin

If the different aquo-ion characteristics of Ca^{++} and Mg^{++} are taken into account, then exchange of the tightly bound divalent cation on actin can be represented by a pseudo-first order competitive model:

$$\mathrm{Ca\text{-}A} \underset{k_{Ca}[\mathrm{Ca}]}{\overset{k_{-Ca}}{\rightleftharpoons}} \mathrm{A} \underset{k_{Mg}[\mathrm{Mg}]}{\overset{k_{-Mg}}{\rightleftharpoons}} \mathrm{Mg\text{-}A}$$

(Equation 1)

---

[1] During the approach to the second plateau ($t_{1/2} \approx 40$ sec), [Ca] is quite low and [Mg] is high so that the apparent rate constant for dissociation is effectively $k \approx k_{-Ca} = 0.017$ sec^{-1}. The situation is slightly different for the approach to the first plateau, which is somewhat faster. Here, with [Mg] = 0, the apparent rate constant for dissociation is $k \approx k_{-Ca} + k_{Ca} \cdot [\mathrm{Ca}] \approx (0.017 + 2 \cdot 10^{7} \cdot 1 \times 10^{-9}) = 0.037$. Thus $t_{1/2}$ would be about 19 sec, in good agreement with the value of 23 sec measured for Fig. 1.

Here Ca-actin (Ca-A) exchanges to Mg-actin (Mg-A) or vice versa through the intermediate divalent cation free actin A. The pseudo-first order character derives from the assumption that the free concentrations of Ca^{++}, [Ca], and Mg^{++}, [Mg], remain approximately constant during the exchange. If it is assumed that A is small, the solution to Equation 1 is a simple exponential time course with apparent rate constant for exchange k_{app}:

$$k_{app} = \frac{k_{-Mg}}{1+\left(\dfrac{k_{Mg}[Mg]}{k_{Ca}[Ca]}\right)} + \frac{k_{-Ca}}{1+\left(\dfrac{k_{Ca}[Ca]}{k_{Mg}[Mg]}\right)}$$

(Equation 2)

(see Gershman et al., 1991, for the full derivation.) At the completion of exchange at free concentrations [Ca] and [Mg], the fraction f_{Ca} of the total actin which will be Ca-actin is:

$$f_{Ca} = \frac{1}{1+\left(\dfrac{K_{Ca}}{K_{Mg}}\cdot\dfrac{[Mg]}{[Ca]}\right)}$$

(Equation 3)

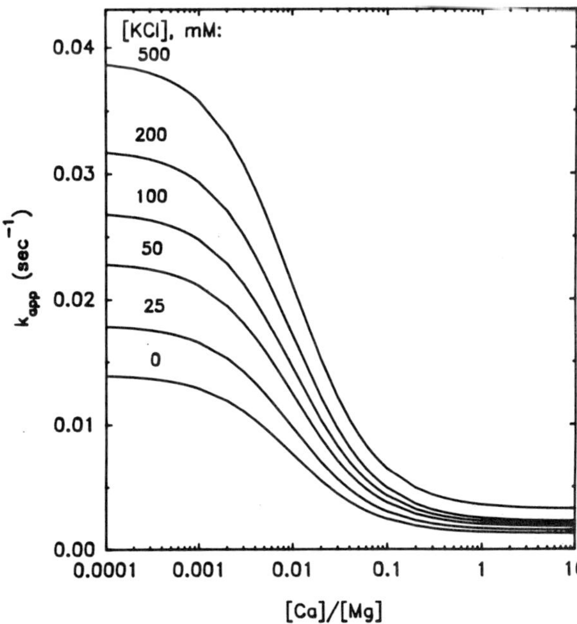

Figure 2. Apparent rate constant k_{app} for divalent cation exchange as a function of free concentration ratio [Ca]/[Mg] at various KCl concentrations. k_{app} is calculated from Equation 2 based on values for k_{-Ca} and k_{-Mg} previously reported (Selden et al., 1989).

In Figure 2, graphs of k_{app} as a function of the ratio [Ca]/[Mg] are shown. (The curves show the results at several KCl concentrations from 0-500 mM.) The curve for 0 KCl, for example, shows that as [Ca]/[Mg] increases, k_{app} decreases smoothly from $k_{-Ca} \approx 0.014$ (at the left) to $k_{-Mg} \approx 0.0012$ (at the right). The midway point of the curve ($k_{app} = \frac{1}{2}(k_{-Ca} + k_{-Mg})$) occurs approximately at [Ca]/[Mg] ≈ 0.01; this reflects the markedly slower association rate constant for Mg^{++} compared with Ca^{++}. The competitive exchange model has been fit to measurements of Ca^{++}/Mg^{++} exchange under various solution conditions with excellent results (Gershman et al., 1991). As noted above, k_{-Ca} and k_{-Mg} increase with pH (Estes et al., 1987) and ionic strength (Selden et al., 1989). The effect of ionic strength on k_{app} can be seen in Figure 2; k_{app} vs [Ca]/[Mg] shifts progressively higher with increasing [KCl] from 0 to 500 mM. k_{-Ca}, k_{-Mg}, and intermediate values of k_{app} are approximately doubled as [KCl] is increased from 0 to the physiologic range (100-200 mM). Thus increased ionic strength weakens high affinity divalent cation binding to actin (Selden et al., 1989) and speeds up the exchange kinetics.

High Affinity Divalent Cation Exchange and Polymerization

As noted earlier, actin is generally prepared and stored as Ca-actin. Mg-actin is easily prepared by incubating the Ca-actin with a Ca-specific chelator such as EGTA or BAPTA, at a concentration twice that of the total Ca^{++} present, and 50-100 μM $MgCl_2$ (see, for example, Gershman et al., 1984). If the cation concentrations have been adjusted so that the concentration ratio [Ca]/[Mg] is less than 0.01, the exchange will be relatively complete in 5 minutes or so. If [Ca]/[Mg] is greater than 0.01, however, the exchange will occur much more slowly (see Figure 2), and even if given additional time, will be incomplete with over 5% contamination with Ca-actin (since $K_{Mg} \approx 5 \cdot K_{Ca}$). Thus some care must be taken in preparing Mg-actin.

For experiments on polymer actin, Ca-actin or Mg-actin can be polymerized with KCl to provide Ca-actin or Mg-actin polymer. However, polymerization of actin has traditionally been induced with KCl and $MgCl_2$ (typically 1-2 mM), or sometimes with $MgCl_2$ alone. In the past, when the affinity of actin for divalent cation was thought to be in the micromolar range, it was assumed that divalent cation exchange was rapid and that polymerization of Ca-actin in the presence of 1-2 mM $MgCl_2$ would yield predominantly Mg-actin polymer. This simply does not happen.

If the tightly-bound divalent cation on actin were to fully equilibrate in solution with free concentrations of [Ca] and [Mg] before polymerization, the fraction of total actin which is Ca-actin would be f_{Ca}, from Equation 3. If we assume that Ca-actin and Mg-actin polymerize with equal rate constants for elongation k_+ (a reasonable approximation; see Selden et al., 1986), and if the starting Ca-actin were polymerized very slowly (but completely) in the presence of divalent cation concentrations [Ca] and [Mg], the proportion of Ca-actin incorporated in the polymer would also be f_{Ca}. With faster polymerization of the initial Ca-actin, the proportion of Ca-actin incorporated in the polymer can only be greater than f_{Ca} as polymerization competes with the Ca^{++}/ Mg^{++} exchange reaction. (It is assumed that polymerization goes to completion and that divalent cation exchange within the polymer does not occur.) Calculating the divalent cation exchange rate constant k_{app} from equation 2 and f_{Ca} from Equation 3, and assuming an apparent pseudo-first order rate constant for polymerization of $(k_+ \cdot m)$, where m is the approximate number concentration of polymers at mid-polymerization, the proportion of Ca-actin in the polymer when Ca-actin is polymerized/exchanged   at divalent cation concentrations [Ca] and [Mg] will be approximately:

$$\frac{[\text{Ca--polymer}]}{[\text{Total polymer}]} = f_{Ca} + \frac{(1-f_{Ca})(k_+ \cdot m)}{(k_+ \cdot m) + k_{app}}$$

(Equation 4)

In Figure 3, Equation 4 is graphed for four divalent cation conditions ($[Ca]/[Mg]$ = 0.001, 0.01, 0.1, 1.0), over a range in $(k_+ \cdot m)$ of 0 to 0.007 sec^{-1} - which covers very slow to rapid polymerizations. The intercepts with the y axis (at the left) reflect the slow polymerization limit (f_{Ca} from Equation 3) for each $[Ca]/[Mg]$ condition. For each of the four conditions, the proportion of Ca-actin incorporated increases with increasing polymerization rate. From the second lowest curve, we see that rapid polymerization (e.g., in 0.1 M KCl, 2 mM MgCl$_2$) of Ca-actin at $[Ca]/[Mg]$ = 0.01 ($[Ca] \approx 20$ μM, $[Mg] \approx 2$ mM, a hundred fold excess of Mg^{++}) would yield polymer containing nearly 40% Ca-actin!

To substantiate these calculations, some simple polymerizations of Ca-actin under various conditions of $[Ca]/[Mg]$ were done in our laboratory, and the proportion of Ca-actin in the resultant polymer was measured. The results are shown in Table 1 and are compared with the calculated percentage of Ca-actin in the polymer based on Equation 4 and Figure 3.   The values for $(k_+ \cdot m)$ were

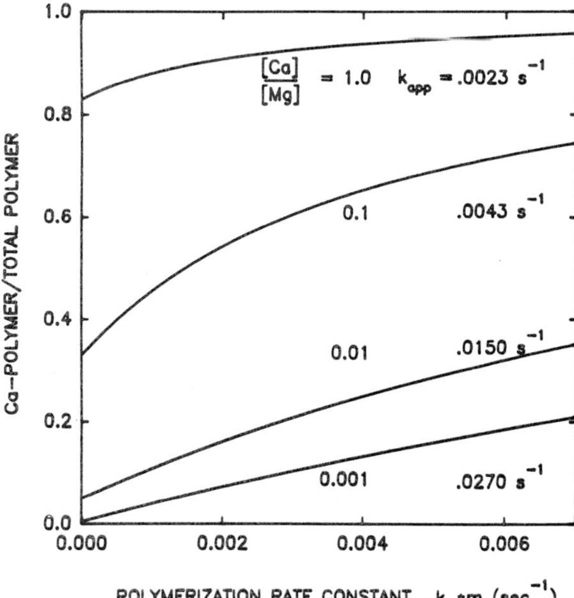

Figure 3. Ca-actin as a proportion of total polymer produced on polymerization of Ca-actin with 0.1 M KCl at several [Ca]/[Mg] ratios. The curves are calculated from Equation 4, using k_{app} values shown, which in turn were calculated from Equation 2 using previously reported values for k_{-Ca} and k_{-Mg} (Selden et al., 1989). The abscissa spans a range for the apparent polymerization rate constant $(k_+ \cdot m)$ of 0 - 0.007 sec^{-1}.

Table 1. Percentage of Ca-actin in the polymer after polymerization of Ca-actin under various conditions.

[KCl] (M)	[CaCl] (mM)	[EGTA] (mM)	$\frac{[Ca]}{[Mg]}$	$(k_+ \cdot m)$ $(sec)^{-1}$	k_{app} $(sec)^{-1}$	Ca-actin Content calculated	measured
0.1	0.20	1	0.0011	0.0047	0.027	10%	15%
0.1	0.02	0	0.0110	0.0035	0.015	17%	21%
0.1	0.20	0	0.1100	0.0022	0.005	32%	34%
0	0.20	1	0.0004	0.0034	0.015	15%	15%
0	0.02	0	0.0110	0.0025	0.006	38%	41%
0	0.20	0	0.1100	0.0025	0.002	60%	66%

Calculated values per legend to Figure 3. All samples contained 10 μM Actin in 5 mM Hepes, 0.2 mM ATP, 0.01% NaN_3 and the indicated $CaCl_2$ and EGTA concentrations. 0 or 0.1 M KCl and 2 mM $MgCl_2$ were added to initiate polymerization. The polymerization was monitored and after complete polymerization, F-actin was separated by centrifugation. Ca^{++} content of the pellet was determined using Quin 2. Mg^{++} content was determined using 8-hydroxyquinoline-5-sulfonic acid at pH 8.0 (Estes et al., 1987) to measure the Mg^{++} content before and after denaturation of the protein. This procedure corrected for secondary binding of Mg^{++} which was minimal for samples containing KCl, but was substantial (≈ 5 Mg^{++} ions per actin molecule) in the absence of KCl. The percent Ca-actin content reported is the average of two determinations with 2 outlier values dropped.

estimated from the half-time $(t_{\frac{1}{2}})$ of polymerization; $(k_+ \cdot m) = 0.693/t_{\frac{1}{2}}$. There is good agreement between the measured values and the calculated values. Note that even if Ca-chelators are added with the Mg^{++} and polymerizing salt, there is a significant proportion of Ca-actin incorporated in the polymer. The only way to be sure of obtaining pure Mg-actin polymer is to exchange the divalent cation at a very low [Ca]/[Mg] ratio before polymerization.

Divalent Cation Binding and Actin Preparation

The somewhat lengthy preparation of actin solutions in the laboratory involves prolonged treatment (polymerization/depolymerization, washing, dialysis, chromatography) with solutions which typically contain ≈ 0.1 mM $CaCl_2$. $MgCl_2$ is generally present only in polymerization buffers, and, as has been shown above, when actin is polymerized rapidly there is little exchange of Mg^{++} into the high affinity binding site. Furthermore, the binding affinity of actin for Ca^{++} is 5 times higher than for Mg^{++}. Thus, during the preparation of actin, Mg^{++} that was originally bound at the high affinity site is replaced by Ca^{++}. This provides an important benefit, as it is generally known that Ca-actin is more stable on storage than is Mg-actin. However, Ca-actin and Mg-actin differ significantly, particularly in their polymerization characteristics and in their abilities to hydrolyze ATP (Estes et al., 1992). Therefore, for studies of actin in its physiologic state, it is important to convert the actin back to the Mg-actin form. This procedure is complicated by several factors: actin binds Ca^{++} better than Mg^{++}; the exchange kinetics are relatively slow; and laboratory reagents frequently are significantly contaminated with Ca^{++}. Thus the conversion of Ca-actin to Mg-actin generally requires the use of chelators and prolonged (5-10 min) exchange times. Lastly, it is prudent to check that the exchange to Mg-actin really is complete by determining the actin-bound cation.

NUCLEOTIDE BINDING TO ACTIN

Background

Each actin molecule binds tightly one nucleotide, which is ATP or ADP under physiological conditions. Many measurements of the nucleotide binding affinity of actin have been made, with little attention to the long known effects of divalent cation on nucleotide binding (see Kinosian et al., 1993, for review). The atomic structure of the actin:DNase I complex recently reported by Kabsch et al. (1990) makes clear the intimate relationship between the tightly bound divalent cation and nucleotide. The bound nucleotide clearly affects divalent cation binding (Selden et al., 1987), and it would be surprising if the reverse were not true.

Several conflicting models for nucleotide binding to actin have been proposed. Frieden and Patane (1988) maintained that only free ATP - not the Ca-ATP complex - can bind to actin, and that there is allosteric control of ADP binding by Ca^{++}. Nowak et al. (1988) proposed that the Ca-ATP complex bound preferentially, but that the Ca^{++} in the complex is not the high affinity divalent cation. Valentin-Ranc and Carlier (1989) showed that the nucleotide-metal complex contains the high affinity divalent cation which is bound to the β and γ phosphates of the actin-bound ATP. This finding was confirmed by the atomic structure determined by Kabsch et al. (1990). Although controversy remains about the mechanism of nucleotide binding to actin, there is some consensus on the structure of the nucleotide-divalent cation binding site: the bound ATP is in a cleft between the two major domains of actin, appearing to hold them together, and the high affinity divalent cation is coordinated by the phosphates of the ATP and the amide groups of the protein.

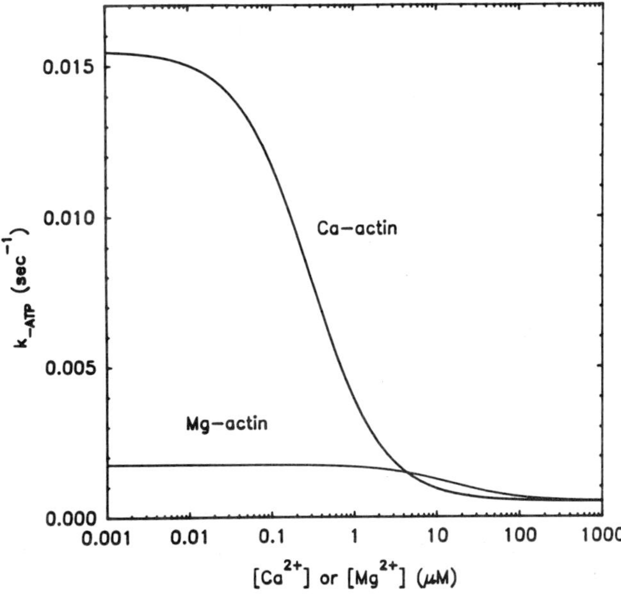

Figure 4. Rate constant k_{-ATP} for dissociation of ATP from Ca-actin and Mg-actin as a function of [Ca] and [Mg]. Curves are calculated from Equation 1 of Kinosian et al. (1993), using constants from Table 1 of the same reference.

Kinetics of High Affinity Nucleotide Binding

The divalent cation concentration clearly regulates high affinity nucleotide dissociation. Figure 4 shows the dependence of ATP dissociation (k_{-ATP}) from Ca-actin and Mg-actin as a function of divalent cation concentration. At low [Ca], the dissociation of ATP from Ca-actin is limited by the dissociation of the high affinity Ca^{++} ($k_{-Ca} \approx 0.015$ sec^{-1}). Presumably the Ca^{++} dissociates first, then the nucleotide dissociates very rapidly (dissociation constant ≈ 6 sec^{-1}) from the divalent cation-free actin. At higher [Ca], the dissociation of ATP from Ca-actin is much slower, and at very high [Ca] the limiting value of k_{-ATP} probably reflects dissociation of the Ca-ATP complex.

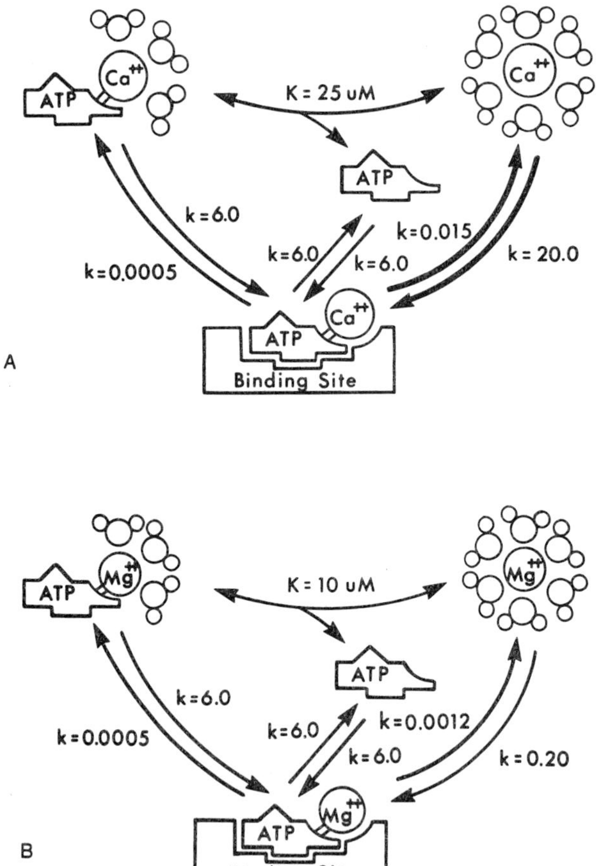

Figure 5. Schematic representation of the binding of Ca^{++} and ATP (A) and Mg^{++} and ATP (B) to actin. K's indicate equilibrium dissociation constants for Ca-ATP and Mg-ATP at pH 7. k's are rate constants in sec^{-1} or M^{-1}sec^{-1} for the indicated reactions. Heavy arrows in (A) indicate reactions with rate constants significantly greater than for the equivalent reactions in (B).

The situation is somewhat different for Mg-actin. At low [Mg], ATP dissociation from Mg-actin is limited by k_{-Mg}, and since k_{-Mg} is much smaller than k_{-Ca}, dissociation of ATP from Mg-actin at low [Mg] is much slower than dissociation of ATP from Ca-actin at low [Ca]. As [Mg] increases, the k_{-ATP} decreases by only a factor of 3. At very high [Ca] or [Mg], k_{-ATP} is about the same for Ca-actin and Mg-actin - suggesting that dissociation of the Ca-ATP complex and the Mg-ATP complex from actin are similar. The divalent cation concentration ranges over which ATP dissociation is modulated are markedly different for Ca-actin and Mg-actin, being about 0.01-1.0 μM for [Ca] and 1.0-100 μm for [Mg]. This difference - a factor of ≈ 100 - results from the different characteristics of the Ca^{++} and Mg^{++} aquo-ions. Note that under physiologic conditions of high [Mg] (≈ 1 mM), nucleotide exchange on actin may be expected to be predominantly via the Mg-nucleotide complex.

Figure 5 shows schematically the interrelationships in the binding of Ca^{++} (A) and Mg^{++} (B) with ATP and actin. The upper portion of each scheme reflects the equilibrium binding of Ca^{++} or Mg^{++} to ATP which differ only slightly in dissociation constant. The remainder of each scheme indicates the two major pathways by which nucleotide binding to actin occurs: by direct binding of the divalent cation-nucleotide complex (on the left), with similar binding kinetics for both Ca^{++}-ATP and Mg^{++}-ATP complexes, and by sequential binding or release of the nucleotide and divalent cation (on the right). It is in this sequential pathway that Ca-actin and Mg-actin differ most; the rate constants for association and dissociation of Ca^{++} are much greater (as denoted by the heavy arrows) than for Mg^{++}. At low divalent cation concentrations, contributions of the two pathways for ATP binding are about equal for Mg-actin, but the sequential pathway greatly predominates for Ca-actin.

Nucleotide Exchange on Actin

If the appropriate kinetic rate constants are known, nucleotide exchange on actin may be described with a pseudo-first order model analogous to the model for high affinity divalent cation exchange (Equation 1):

$$MN_1A \xrightleftharpoons[k_{+N1}[N_{1tot}]]{k_{-N1}} A \xrightleftharpoons[k_{+N2}[N_{2tot}]]{k_{-N2}} MN_2A$$

$$\text{(Equation 5)}$$

Here N_1 and N_2 are two nucleotides at concentrations $[N_{1tot}]$ and $[N_{2tot}]$ competing in solution to produce the metal-nucleotide-actin complexes MN_1A and MN_2A via the intermediate divalent cation-free/nucleotide-free actin, A (see Kinosian et al., 1993). Only one divalent cation, M, is present, and the apparent association rate constants k_{+N1} and k_{+N2} and dissociation rate constants k_{-N1} and k_{-N2} may be expressed as functions of the concentration of M and the appropriate rate constants from a scheme such as that in Figure 5A or 5B (Kinosian et al., 1993). The solution for this model is analogous to that for divalent cation exchange (Equations 2 and 3):

$$k_{ex} = \cfrac{k_{-N1}}{1+\left(\cfrac{k_{+N1}[N_{1tot}]}{k_{+N2}[N_{2tot}]}\right)} + \cfrac{k_{-N2}}{1+\left(\cfrac{k_{+N2}[N_{2tot}]}{k_{+N1}[N_{1tot}]}\right)}$$

(Equation 6)

$$\frac{[MN_1A]}{[actin_{tot}]} = \cfrac{1}{1+\left(\cfrac{K_D^{N1}}{K_D^{N2}}\cdot\cfrac{[N_{2tot}]}{[N_{1tot}]}\right)}$$

(Equation 7)

k_{ex} from Equation 6 is the apparent rate constant for nucleotide exchange on actin. The extent of exchange is expressed in Equation 7 as the fraction of total actin ($actin_{tot}$) containing nucleotide N_1. Here, $K_D^{N1} = k_{-N1}/k_{+N1}$ and similarly, $K_D^{N2} = k_{-N2}/k_{+N2}$ (see Kinosian et al. 1993, for a more complete description).

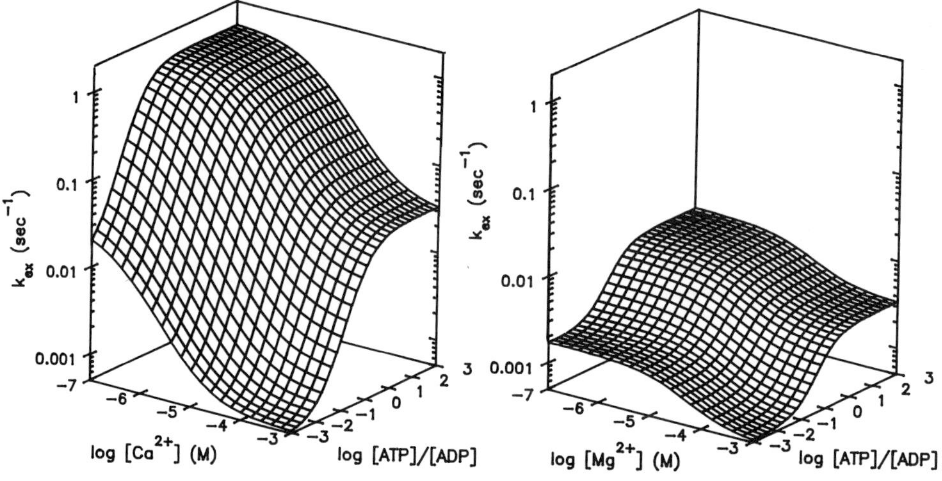

Figure 6. Nucleotide exchange rates on actin. Theoretical plots of the ATP/ADP exchange rate constant k_{ex} as a function of [ATP]/[ADP] and of divalent cation concentration for Ca-actin (left) and Mg-actin (right). Note that all scales are logarithmic. The plots were generated from Equation 6 using the appropriate constants from Kinosian et al. (1993), from which the figure is reprinted with permission.

Figure 6 shows theoretical plots of the calculated rates of exchange (apparent rate constant k_{ex}) of ATP and ADP on Ca-actin (left) and Mg-actin (right) as functions of the divalent cation concentration [Ca] or [Mg] and of the ratio of nucleotide concentrations [ATP]/[ADP] (this figure and the data on which it is based are from Kinosian et al., 1993). Note the markedly wider range in k_{ex} and the markedly faster rates of exchange at low divalent cation concentrations for Ca-actin compared with Mg-actin. Figure 6 also indicates that under physiologic conditions ([Mg] $\approx$ 1 mM, high ratio of [ATP]/[ADP]), the exchange of ATP for

ADP on Mg-ADP actin will be relatively slow with $k_{ex} \approx .003$ sec^{-1}. This may well have implications for the *in vivo* recycling of dissociated ADP-actin during cytoskeletal remodeling; without some assistance, such as by actin-associated proteins (e.g. Goldschmidt-Clermont et al., 1992), nucleotide exchange and reutilization of these monomers might be expected to be quite slow.

Preparation of ADP-actin

An understanding of the high affinity divalent cation binding characteristics of actin is useful when one attempts to prepare ADP-actin. The relative affinity of actin for Ca^{++} vs Mg^{++} is reversed for ADP-actin compared with ATP-actin; and Ca^{++} binds about 50-fold weaker to ADP-actin compared with ATP-actin (Selden et al., 1987). Ca-ADP-actin is more subject to denaturation than Ca-ATP-actin, unless care is taken to keep the solvent Ca^{++} and ADP concentrations high. Furthermore, it is difficult to prepare Ca-ADP-actin since ADP-actin binds Mg^{++} 20-fold stronger than it binds Ca^{++} so that contaminant Mg^{++} is easily taken up by the ADP-actin. The use of hexokinase/glucose to prepare ADP-actin (initially reported by Pollard, 1984) is most applicable for the preparation of Mg-ADP-actin, since the hexokinase requires Mg^{++} for activity. The trick then is to replace the bound Mg^{++} with Ca^{++}; this can be done using the mass action effect of Ca^{++}-saturated Chelex® (Gershman et al., 1989). Unfortunately, although Ca^{++}/Mg^{++} exchange is fairly complete, there is some loss of protein with the Chelex® procedure. Alternatively, Ca-ADP-actin can be prepared more directly by polymerizing Ca-ATP-actin with 20 mM KCL, 0.5 mM ADP, and 0.2 mM $CaCl_2$ (after removing excess solvent ATP with Dowex AG-1), using two or three brief sonications to promote hydrolysis of the remaining actin-bound ATP, then dialyzing at 4°C against 0.5 mM ADP, 0.2 mM $CaCl_2$, in 5 mM Hepes, pH 7.0, to depolymerize the actin. This procedure is described more completely in Kinosian et al. (1993). Since Mg^{++} exposure is avoided with this procedure, exchange to Ca-ADP-actin is quite complete.

SUMMARY AND CONCLUSIONS

At this point, it may be worthwhile to list, in summary form, the important aspects of divalent cation and nucleotide binding to actin that have been reviewed here:

1) High affinity divalent cation binding to actin is very tight, with equilibrium dissociation constant $K_{Ca} \approx 1$ nM and $K_{Mg} \approx 5$ nM at pH 7.0.

2) The binding kinetics of Ca^{++} are diffusion limited. Dissociation is slow, with $k_{-Ca} \approx 0.015$ sec at pH 7.0 (and low ionic strength).

3) The binding kinetics of Mg^{++} are limited by the characteristics of the Mg^{++} aquo-ion and are much slower than for Ca^{++}; $k_{-Mg} \approx 0.0012$ at pH 7.0.

4) Increase in pH or ionic strength weakens divalent cation binding at the high affinity site, primarily by increasing k_{-Ca} and k_{-Mg}.

5) Exchange of Mg^{++} for Ca^{++} (or vice versa) at the high affinity site is by a competitive pseudo-first order process with an apparent rate constant (k_{app}) intermediate between k_{-Ca} and k_{-Mg} and dependent upon the cation concentration ratio [Ca]/[Mg] present.

6) High affinity ATP binding is modulated by the high affinity divalent cation. The cation concentration range over which this modulation occurs is about

100-fold higher for Mg^{++} than for Ca^{++}, again because of the different characteristics of the Mg^{++} and Ca^{++} aquo-ions.

7) At low divalent cation concentrations, ATP dissociation from actin is limited by dissociation of the tightly-bound divalent cation.

8) At high divalent cation concentrations, ATP dissociation probably occurs via dissociation of the divalent cation-nucleotide complex and is quite slow, with dissociation rate constant ≈ 0.0005 sec^{-1}.

9) Competitive nucleotide exchange on actin may be described by a pseudo-first order model analogous to that for divalent cation exchange. The pseudo-first order rate constants depend upon the divalent cation concentration. The overall nucleotide exchange rate constant k_{ex} depends upon these constants and the solution nucleotide concentration ratio, e.g. [ATP]/[ADP].

The following circumstances develop from the characteristics of the high affinity binding of divalent cation and nucleotide to actin:

1) The standard methods for actin preparation convert *in vivo* Mg-actin into Ca-actin.

2) Converting Ca-actin back to Mg-actin is not easy. A very low ratio of [Ca]/[Mg] is necessary, which usually requires the use of Ca-chelators, and a long time (5-10 min) must be allowed for complete exchange.

3) When Ca-actin is polymerized with $MgCl_2$, even at high $MgCl_2$ concentration with Ca-chelator simultaneously added, the polymer produced will be significantly contaminated with Ca-actin.

4) Ca-ADP-actin is hard to prepare, since Ca^{++} binds very weakly to ADP-actin and thus denaturation is accelerated. Use of Ca-saturated Chelex® to exchange Mg-ADP-actin to Ca-ADP-actin may be helpful, or Ca-ADP-actin may be prepared from polymerized Ca-actin (Kinosian et al., 1993).

5) At high divalent cation concentrations, nucleotide exchange is very slow and at pH 7.0 complete exchange may require hours.

Many problems in the biophysics and biochemistry of actin remain incompletely answered. What is the role of ATP-hydrolysis in actin polymerization? Does phosphate (Pi) have a significant *in vivo* role? How do ADP-Pi-actin or ATP-actin caps on actin filaments work? ADP/ATP exchange on actin is slow under physiological conditions; does the cell accelerate this? And how? A better understanding of divalent cation and nucleotide binding to actin may help solve these and other interesting problems about actin.

ACKNOWLEDGEMENT

The authors acknowledge the excellent secretarial assistance of Marie Strouse in preparation of this manuscript. This work was supported by the Department of Veterans Affairs and National Institutes of Health Grant GM 32007.

REFERENCES

Diebler, H., M. Eigen, G. Ilgenfritz, G. Maass, and R. Winkler. 1969. Kinetics and mechanism of reactions of main group metal ions with biological carriers. *Pure Appl. Chem.* 20:93-115.

Estes, J.E. and C. Moos. 1969. Effect of Bound-Nucleotide Substitution on the Properties of F-Actin. *Arch. Biochem. Biophys.* 132:388-396.

Estes, J.E., L.A. Selden, and L.C. Gershman. 1987. Tight Binding of Divalent Cations to Monomeric Actin. *J. Biol. Chem.* 262:4952-4957.

Estes, J.E., L.A. Selden, H.J. Kinosian, and L.C. Gershman. 1992. Tightly-bound divalent cation of actin. *J. Musc. Res. Cell Mot.* 13:272-284.

Frieden, C., D. Lieberman, and Helen R. Gilbert. 1980. A Fluorescent Probe for Conformational Changes in Skeletal Muscle G-Actin. *J. Biol. Chem.* 255:8991-8993.

Frieden, C. 1982. The Mg-induced Conformational Change in Rabbit Skeletal Muscle G-actin. *J. Biol. Chem.* 257:2882-2886.

Frieden, C. and K. Patane. 1988. Mechanism for Nucleotide Exchange in Monomeric Actin. *Bio chemistry* 27:3812-3820.

Gershman, L.C., L.A. Selden, and J.E. Estes. 1986. High Affinity Binding of Divalent Cation to Actin Monomer is Much Stronger than Previously Reported. *Biochem. Biophys. Res. Comm.* 135:607-614.

Gershman, L.C., L.A. Selden, H.J. Kinosian, and J.E. Estes. 1989. Preparation and polymerization properties of monomeric ADP-Actin. *Biochem. Biophys. Acta* 995:109-115.

Gershman, L.C., L.A. Selden, and J.E. Estes. 1991. High Affinity Divalent Cation Exchange on Actin. Association rate measurements support the simple competitive model. *J. Biol. Chem.* 266:76-82.

Goldschmidt-Clermont, P.J., M.I. Furman, D. Wachsstock, D. Safer, V.T. Nachmias, and T.D. Pollard. 1992. The Control of Actin Nucleotide Exchange by ThymosinBeta4 and Profilin. A Potential Regulatory Mechanism for Actin Polymerization in Cells. *Mol. Biol. Cell* 3:1015-1024.

Kabsch, W., H.G. Mannherz, D. Suck, E.F. Pai, and K.C. Holmes. 1990. Atomic structure of the actin: DNase I complex. *Nature* 347:37-44.

Kasai, M. and F. Oosawa. 1969. Behavior of Divalent Cations and Nucleotides Bound to F-actin. *Biochem. Biophys. Acta* 172:300-310.

Kinosian, H.J., L.A. Selden, J.E. Estes, and L.C. Gershman. 1993. Nucleotide Binding to Actin: Cation dependence of nucleotide dissociation and exchange rates. *J. Biol. Chem.* 268: 8683-8691.

Kitazawa, T., H. Shuman, and A.P. Somlyo. 1982. Calcium and magnesium binding to thin and thich filaments in skinned muscle fibres:electron probe analysis. *J. Musc. Res. Cell Mot.* 3:437-454.

Konno, K. and Manuel F. Morales. 1985. Factors in G-actin conformation. *Proc. Natl. Acad. Sci.* 82:7904-7908.

Martonosi, A., C.M. Molino, and J. Gergely. 1964. The Binding of Divalent Cations to Actin. *J. Biol. Chem.* 239:1057-1064.

Maruyama, K. and J. Gergely. 1961. Removal of the bound calcium of G-actin by ethylenediamine tetraacetate (EDTA). *Biochem. Biophys. Res. Commun.* 6:245-249.

Newman, J., K.S. Zaner, K.L. Schick, L.C. Gershman, L.A. Selden, H.J. Kinosian, J.L. Travis, and J.E. Estes. 1993. Nucleotide exchange and rheometric studies with F-actin prepared from ATP- or ADP-monomeric actin. *Biophys. J.* 64:1559-1566.

Nowak, E., H. Strzelecka-Golaszewska, and R. Goody. 1988. Kinetics of nucleotide and metal ion interaction with G-actin. *Biochemistry* 27: 1785-1792.

Pollard, T.D. 1984. Polymerization of ADP-actin. *J. Cell Biol.* 99:769-777.

Selden, L.A., L.C. Gershman, and J.E. Estes. 1986. A kinetic comparison between Mg-actin and Ca-actin. *J. Musc. Res. Cell Mot.* 7:215-224.

Selden, L.A., L.C. Gershman, H.J. Kinosian, and J.E. Estes. 1987. Conversion of ATP-actin to ADP-actin reverses the affinity of monomeric actin for Ca vs Mg. *FEBS Lett.* 217:89-93.

Selden, L.A., J.E. Estes, and L.C. Gershman. 1989. High Affinity Divalent Cation Binding to Actin Effect of Low Affinity Salt Binding. *J. Biol. Chem.* 264:9271-9277.

Tsien, R.Y., T. Pozzan, and T.J. Rink. 1982. Calcium Homeostasis in Intact Lymphocytes: Cytoplas mic Free Calcium Monitored With a New, Intracellularly Trapped Fluorescent Indicator. *J. Cell Biol.* 94:325-334.

Valentin-Ranc, C. and M.-F. Carlier. 1989. Evidence for the direct interaction between tightly bound divalent metal ion and ATP on actin: binding of the isomers of beta and gamma-bidendate CrATP to actin. *J. Biol. Chem.* 264:20871-20880.

Weber, A., R. Herz, and I. Reiss. 1969. The role of magnesium in the relaxation of myofibrils. *Biochemistry* 8:2266-2271.

INFLUENCE OF THE HIGH AFFINITY DIVALENT CATION ON ACTIN TRYPTOPHAN FLUORESCENCE

Lynn A. Selden, Henry J. Kinosian, James E. Estes,
and Lewis C. Gershman

Research and Medical Services
Stratton VA Medical Center
Albany, New York 12208
and
Departments of Medicine and Physiology and Cell Biology
Albany Medical College
Albany, New York 12208

INTRODUCTION

The ability of actin to form stable polymers is important to many cellular processes. Actin polymerization has been measured by a wide variety of methods. Kerwar and Lehrer (1) were the first to demonstrate tryptophan fluorescence changes upon denaturation and polymerization of actin. We and others have made use of this work in assessing actin denaturation (2-4); however, intrinsic fluorescence has not been exploited as a general means of following actin polymerization. Instead, fluorescent probes such as pyrene attached near the C-terminal end of the actin molecule have been the primary tools for monitoring actin polymerization. This approach has yielded a wealth of information, but the possibility that labeled actin does not exactly reflect the characteristics of native actin always remains.

In recent years it has been widely recognized that the properties of actin containing Mg^{2+} at the high affinity divalent cation binding site (Mg-actin) are quite different from those of actin containing Ca^{2+} at this site (Ca-actin). Interestingly, most of the data concerning divalent cation exchange kinetics has come from studies using actin fluorescently labeled at the same site as is used for polymerization studies. This suggests that the environment in this region of the actin molecule changes both during polymer formation and on exchange of the divalent cation. Since actin contains 2 tryptophan residues in close proximity to the C-terminal (5), the tryptophan fluorescence of actin might be affected by the tightly bound divalent cation. In this study, we confirm that changes in actin tryptophan fluorescence are useful in monitoring actin polymerization, and we show the effects of the tightly bound divalent cation on the tryptophan fluorescence characteristics.

Actin: Biophysics, Biochemistry, and Cell Biology
Edited by J.E. Estes and P.J. Higgins, Plenum Press, New York, 1994

MATERIALS AND METHODS

Actin was extracted from rabbit muscle acetone powder and purified by previously published procedures (6). Fluorescence data were collected with an SPF 500-C spectrofluorometer (SLM Instruments) using 10 X 10 mm quartz cuvettes. Tryptophan fluorescence was excited at 300 nm and intensity changes in the emission fluorescence were measured at 335 nm. Light scattering was observed at 300 nm. ATP and Quin2 were purchased from Sigma Chemical Company. Monomeric Ca-actin was in G-buffer: 5mM Hepes, 0.2mM ATP, 0.02mM $CaCl_2$, and 1.5mM NaN_3, pH 7.0. Mg-actin monomer was prepared just prior to use by incubation of Ca-actin in G-buffer with 100 μM $MgCl_2$ and 100μM EGTA for at least 6 min. Polymerization buffer (F-buffer), was prepared by addition of 0.15M KCl to G-buffer. Divalent cation free actin (DCF-actin) was prepared by incubation of Ca-actin for 10 minutes in the presence of 2mM EDTA, pH 7.0, 5°C or at 25°C with 2mM EDTA and 2mM ATP.

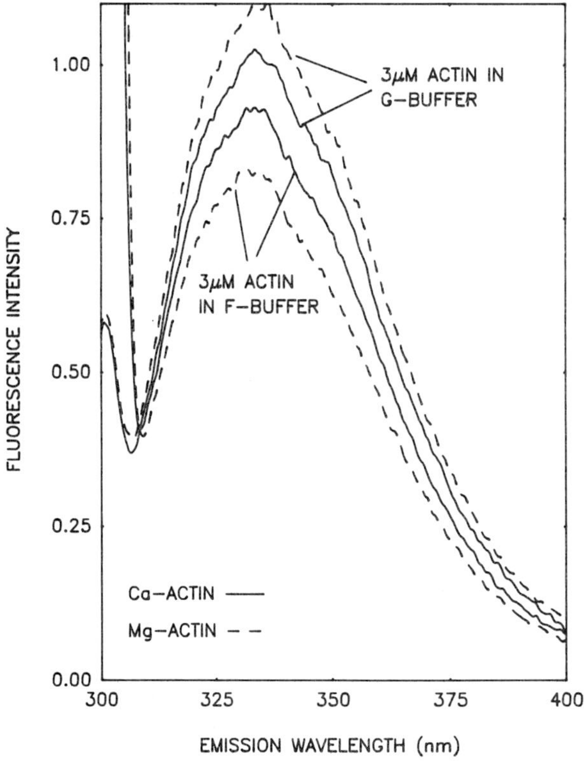

Figure 1. Emission spectra of actin as a function of tightly bound divalent cation. Ca-actin (——) and Mg-actin (---), 3μM, in G-buffer (upper spectra) or polymerized in F-buffer (lower spectra) were prepared and spectra determined as described in the Materials and Methods.

RESULTS AND DISCUSSION

Figure 1 shows tryptophan intrinsic fluorescence emission spectra of Ca-actin (solid lines) and Mg-actin (dashed lines) in the monomeric state in G-buffer (upper two curves) and in the polymeric state in F-buffer (lower two curves). No spectral shifts are evident, but there are clearly fluorescence intensity differences between Ca-actin and Mg-actin. Monomeric Mg-actin has a tryptophan fluorescence intensity 10% higher than that of Ca-actin monomer. On polymerization of Mg-actin and Ca-actin, fluorescence intensity is reduced by approximately 25% and 10% respectively. The difference between the fluorescence intensities of Mg-actin and Ca-actin polymers is in part due to a difference in critical concentrations (6), but this does not account for the entire difference. From Figure 1 it is clear that the fluorescence decrease on polymerization of Mg-actin is much larger than that for Ca-actin. Light scattering intensities similar to those caused by polymeric actin were simulated by addition of soluble dextran to monomeric actin solutions; these samples showed no fluorescence intensity decrease due to scattering of the excitation light by dextran "equivalent" to 5μM actin polymer and only minimal effects up to the equivalent of 20μM polymer. Thus, the decrease in tryptophan fluorescence intensity with polymerization is not caused by loss of excitation due to scattering. The spectra shown in Figure 1 are similar to those observed by Kerwar and Lehrer (1) and are characteristic of proteins in which the indole ring of tryptophan lies inside the protein rather than on the surface (7).

In Figure 2 the time course of monomeric actin bound divalent cation exchange is compared with the time course of the tryptophan fluorescence intensity change. The temporal relationship between the exchange reaction as monitored by Quin2 and the tryptophan fluorescence change under similar conditions demonstrates that the intrinsic tryptophan fluorescence intensity change occurs as a consequence of either Ca^{2+} removal or Mg^{2+} binding. Previously, we have shown

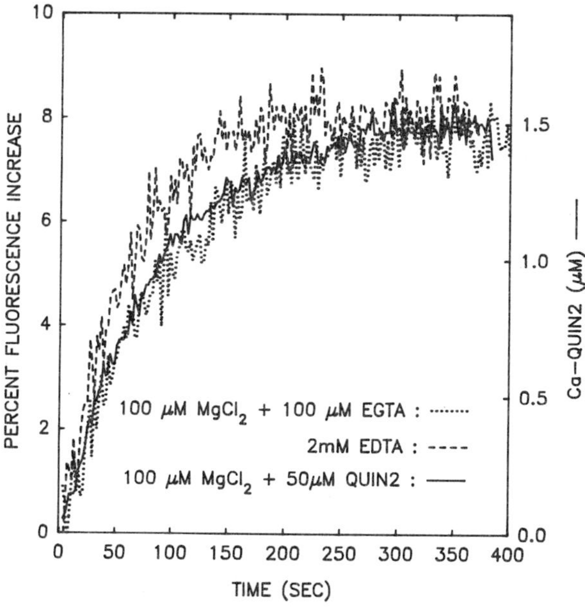

Figure 2. Time course of divalent cation exchange and Ca^{2+} removal as monitored by Quin2 and actin tryptophan fluorescence. At time zero, the indicated amount of chelator and $MgCl_2$ was added and the emission intensity of either Quin2 or tryptophan was monitored as described in Materials and Methods.

that the fluorescence of 1,5-I-AEDANS-actin increases upon release of bound Ca^{2+} rather than upon binding of Mg^{2+}. Thus, the fluorescence characteristics of 1,5-I-AEDANS-DCF-actin are similar to those of 1,5-I-AEDANS-Mg-actin. Addition of 2 mM EDTA to Ca-actin in Figure 2 shows that the fluorescence intensity increase during formation of DCF-actin is similar to that of Mg-actin and is in good agreement with the 1,5-I-AEDANS-actin studies (8). The slightly faster fluorescence change in the presence of 2 mM EDTA most likely reflects the effect of ionic strength on the exchange reaction due to the EDTA and higher ATP concentration (2mM) used in this sample (9). Figure 3 shows the fluorescence emission spectra of the samples from Figure 2 and, for comparison, the fluorescence emission spectrum of denatured actin. Note, as first observed by Kerwar and Lehrer (1), that denaturation causes a spectral shift which has been useful in assessing denaturation of actin (2-4). ATP concentrations from 20 - 2000 μM had

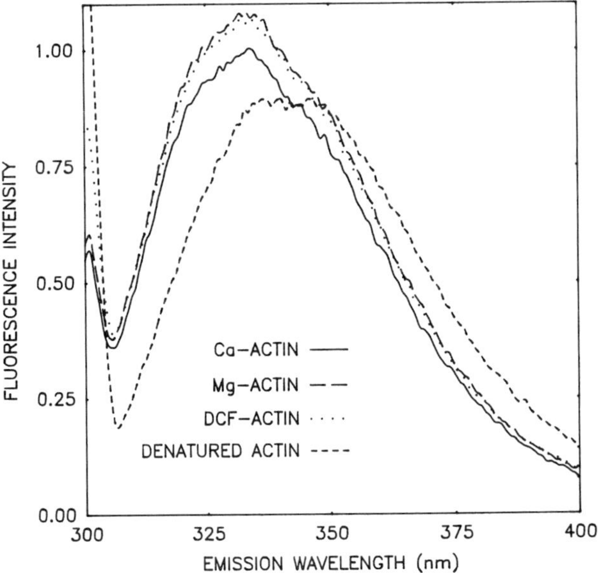

Figure 3. Fluorescence emission spectra of Ca-actin, Mg-actin, DCF-actin, and denatured actin. Ca-actin, Mg-actin, and DCF-actin were prepared as described in Materials and Methods. Denatured actin was prepared by extended incubation of DCF-actin at 25°C in the presence of low (20μM) ATP concentration.

no effect on the intrinsic fluorescence of Ca-actin, Mg-actin or DCF-actin, nor was there any difference between ADP-actin and ATP-actin (data not shown).

Given the difference in tryptophan fluorescence between monomer and polymer actin, one should be able to follow the formation of polymer with time. The data shown in Figure 4 verify that the tryptophan fluorescence decrease upon polymer formation correlates well with the light scattering increase measured on the same sample. Figure 5 shows the change in tryptophan fluorescence upon polymerization as a function of actin concentration for Ca-actin and Mg-actin. From a number of experiments we have determined that the fluorescence intensity decreases 27 ± 2 % for Mg-actin (12 determinations on 6 separate actin preparations), while Ca-actin undergoes only a 10% decrease in tryptophan fluorescence intensity. This figure demonstrates the utility of the fluorescence intensity change in determining the critical concentration of native actin. Also shown in Figure 5, (◇) are the initial polymerization rates for different concentrations of Mg-actin

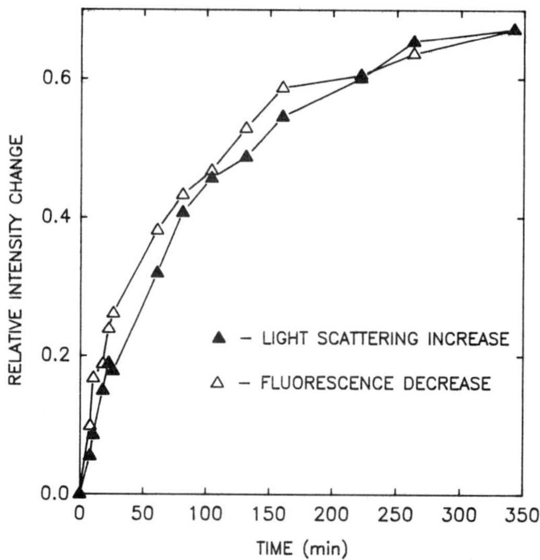

Figure 4. Temporal relationship of actin tryptophan fluorescence decrease and the increase in light scattering associated with formation of actin polymer. Ca-actin in G-buffer at 6μM was polymerized by addition of 0.1M KCl.

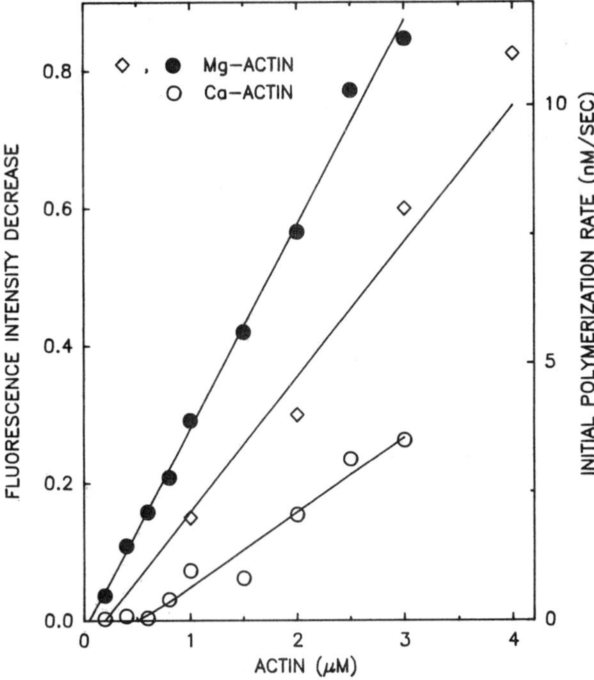

Figure 5. Determination of the critical concentration of Mg-actin, ●, and Ca-actin, ○, by the difference between monomer and polymer tryptophan fluorescence intensity. Samples of Mg-actin and Ca-actin monomer were incubated overnight with and without 0.15M KCl and the fluorescence change determined. Critical concentration determination by initial polymerization rate method, ◇. Sonicated polymer was added to Mg-actin in the presence of 0.15M KCl and the change in fluorescence intensity monitored. Fluorescence change was converted to polymer concentration using the slope of the Mg-actin critical concentration curve.

to which nuclei have been added. Plots such as these are useful in determining the rate constants of polymerization as well as the critical concentration of actin (9).

Figure 6 shows the application of the actin tryptophan fluorescence to the determination of polymer nucleus size. The half times of polymerization for 1-5 µM Mg-actin are determined from the polymerization time course data in Figure 6A and are plotted as a function of actin concentration in Figure 6B. From a curve fit to this data, a nucleus size of 3.2 monomers can be calculated, in good agreement with the values in the literature (10).

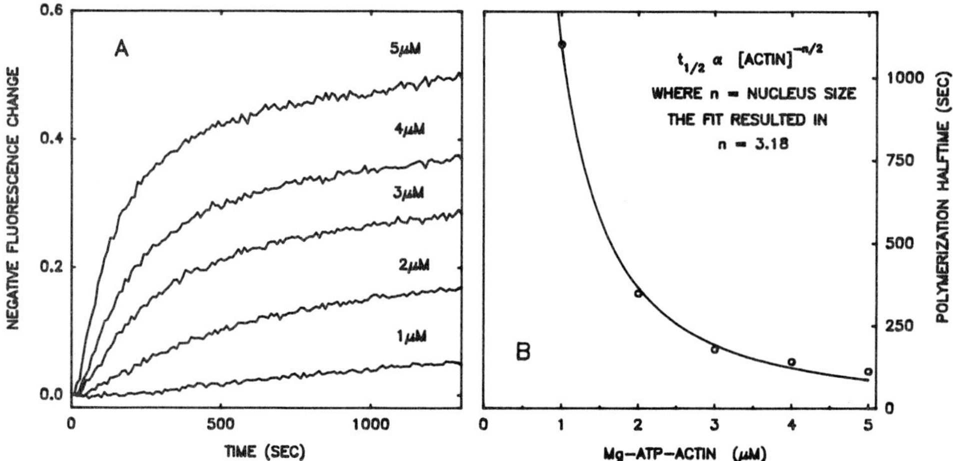

Figure 6. Monitoring polymerization and determining nucleus size for Mg-actin using actin tryptophan fluorescence. A: Mg-actin, 1-5µM in G-buffer was polymerized by addition of 0.15M KCl, and the change in fluorescence was monitored. B: The polymerization half time as determined in A was plotted as a function of concentration and the data were fit by the equation shown in the figure.

SUMMARY

This study demonstrates that the intrinsic tryptophan fluorescence of actin provides an effective and convenient way of measuring divalent cation exchange and polymerization of native actin. In the measurement of divalent cation exchange, this method is as sensitive as those previously used (8) and provides further evidence of the importance of the tightly bound divalent cation to the properties of actin. In monitoring polymerization, this method cannot compete with the sensitivity of the commonly used pyrene-actin fluorescence. However, some proteins (e.g., profilin) and other agents (e.g., cytochalasin D) that bind to actin are affected by the presence of fluorescent labels, making actin intrinsic fluorescence potentially useful in investigating the interaction of these agents with actin, and in validating data obtained using labeled actin. Our laboratory routinely checks the quality of our native actin preparations using this technique and, more recently, we have used actin tryptophan fluorescence to monitor the nucleating and severing effects of gelsolin on actin. The simplicity of the technique is most appealing, and we expect that a variety of innovative and routine uses will be developed.

ACKNOWLEDGEMENT

The authors acknowledge the excellent secretarial assistance of Marie Strouse in the preparation of this manuscript. This work was supported by the Department of Veterans Affairs and National Institutes of Health grant GM 32007.

REFERENCES

1. G. Kerwar and S.S. Lehrer. Intrinsic fluorescence of actin. *Biochemistry* 11:1211-1217 (1972).
2. L.C. Gershman, J.E. Estes, and L.A. Selden, L.A. Polymerization characteristics of divalent cation-free actin. *Ann. N.Y. Acad. Sci.* 259:264-266 (1988).
3. A. Bertazzon, G.H. Tian, A. Lamblin, and T.Y. Tsong. Enthalpic and entropic contributions to actin stability: Calorimetry, circular dichroism, and fluorescence study and effects of calcium. *Biochemistry* 29:291-298 (1990).
4. H.J. Kinosian, L.A. Selden, J.E. Estes, and L.C. Gershman. Nucleotide binding to actin: Cation dependence of nucleotide dissociation and exchange rates. *J. Biol. Chem.* 268:8683-8691 (1993).
5. W. Kabsch, H.G. Mannherz, D. Suck, E.F. Pai, and K.C. Holmes. Atomic structure of the actin: DNase I complex. *Nature* 347:37-44 (1990).
6. L.A. Selden, L.C. Gershman, and J.E. Estes, J.E. A kinetic comparison between Mg-actin and Ca-actin. *J. Musc. Res. Cell Motility* 7:215-224 (1986).
7. E.A. Permyakov. "Luminescent Spectroscopy of Proteins," CRC Press, Florida (1992).
8. J.E. Estes, L.A. Selden, and L.C. Gershman, L.C. Tight-binding of divalent cations to actin: Binding kinetics support a simplified model. *J. Biol. Chem.* 262:4952-4957 (1987).
9. L.A. Selden, J.E. Estes, and L.C. Gershman. High affinity divalent cation binding to actin: Effect of low affinity salt binding. *J. Biol. Chem.* 264:9271-9277 (1989).
10. F. Oosawa. Macromolecular assembly of actin, *in*: "Muscle and Nonmuscle Motility," A. Stracher, ed., Academic Press, New York (1983).

C-TERMINUS ON ACTIN: SPECTROSCOPIC AND IMMUNOCHEMICAL EXAMINATION OF ITS ROLE IN ACTOMYOSIN INTERACTIONS

Anh M. Duong and Emil Reisler

Department of Chemistry and Biochemistry
and the Molecular Biology Institute
University of California
Los Angeles, CA 90024

The understanding of force generation during cyclic interactions of myosin with actin requires a detailed description of actomyosin interface and its evolution during ATP hydrolysis. Until now, driven by the availability of mutants[1-4] and atomic resolution structure of actin[5], the mapping of actomyosin interface has focused on the determination of myosin binding sites on actin. Two proximal areas on actin, in the N- and C-terminal regions of this protein, have been implicated by structural considerations[5,6] and biochemical[7,8], immunochemical[9-13], NMR[14], and mutagenic studies[1-3,15] in the binding of myosin heads (S-1). Immunochemical[10] and mutagenic approaches[16] suggested also the involvement of residues 91-103 on actin in actomyosin interactions. While the previous studies produced a general agreement on the important contribution of N-terminal acidic residues 1-4 on actin to the activation of the myosin ATPase activity and the motility of actin filaments, the actual role of actin's C-terminal residues in actomyosin interactions has not been assessed. An interesting approach to this task, which was employed also in the work on LC-2 myosin light chains[17], was taken by Labbe et al.[10] These authors modified the penultimate cysteine residue on actin, Cys-374, with N-iodoacetyl-N'-(5-sulpho-1-naphtyl)ethylenediamine (1,5- IAEDANS) and showed in solid phase immunochemical assays (ELISA) that S-1 and antidansyl antibodies competed with each other for the binding to the modified cysteine.

The implied involvement of Cys-374 on actin in myosin binding would not be without a precedence. The same residue on actin was shown to bind to profilin and could be cross-linked to caldesmon[19], gelsolin[20], and, in the case of monomeric actin, to myosin heads as well[21]. Yet, the mobile nature of actin's C-terminus calls for some caution in the interpretation of cross-linking and solid-phase immunochemical studies. Indeed, a direct binding of the last few C-terminal residues on actin to myosin is not indicated by experiments with tryptically truncated actin. Proteolytic removal of the last two[22] or three[23] C-terminal residues on actin, including Cys-374, did not alter much the actin-activated ATPase activity of myosin despite some reported changes in the morphology of actin filaments[24].

In this work we have examined the environment of Cys-374 on actin by attaching three different fluorescent probes to this residue. The interaction of myosin with the probe region on actin was assessed in collisional quenching experiments and by using antiprobe antibodies. Our results show probe-dominated changes in the environment of Cys-374 and do not reveal any direct interaction between S-1 and the modified Cys-374 region on actin. Moreover, we did not detect any significant structural coupling between the functionally important N-terminal residues on actin and its C-terminus.

MATERIALS AND METHODS

Reagents

N-dansylaziridine (DAZ) was a product of Pierce Chem. Co. (Rockford, IL). Iodoacetamidofluorescein (Fl) was purchased from Molecular Probes (Eugene, OR). N-iodoacetyl-N'-(5-sulpho-1-naphthyl) ethylenediamine (1,5-IAEDANS), nitromethane, and immunochemicals were from Sigma Chem. Co. (St. Louis, MO). Protein A conjugated to Sepharose CL4B was from Pharmacia (Piscataway, NJ).

Proteins and Antibodies

Rabbit myosin subfragment 1 (S-1) and actin were prepared according to the methods of Weeds and Pope[25] and Spudich and Watt[26], respectively. The monoclonal anti-dansyl (anti-DAZ) IgG was a generous gift from Dr. V. Oi. The F_{ab} fragment of affinity purified polyclonal antibodies against residues 1-7 on α-skeletal rabbit actin (F_{ab} (1-7)) was a gift from Dr. G. DasGupta. Anti-fluorescein antibodies (anti-Fl) were prepared according to Lopatin and Voss[27]. The IgG fraction of the antiserum was prepared by chromatography on a protein A column[11]. Between 6 and 13% of IgG was specific for fluorescein as determined from quenching titrations of fluorescein with the IgG.

Modification of Actin

Cysteine-374 on actin was modified with 1,5 - IAEDANS, N-dansylaziridine and iodoacetamidofluorescein. The labeling with IAEDANS and fluorescein was carried out at a 20-fold molar excess of reagent over protein[28]. The dansylaziridine modification was done at a 40-fold molar ratio of reagent to protein as previously described[29]. The labeling stoichiometry, which was determined spectrophotometrically, was typically between 0.50 and 0.75 mol of probe per mol of actin.

Fluorescence Measurements

Steady-state fluorescence intensity measurements in the presence of acrylamide or nitromethane as quenchers were performed as previously described[30]. Small volumes of quencher were added sequentially to actin in G-actin buffer (0.2mM $CaCl_2$, 20 and

$30\mu M$ ATP, and 5.0mM Tris-HCl, pH 7.6). Solutions of F-actin and acto-S-1 contained also 2.0mM $MgCl_2$. All measurements were done in a Spex Fluorolog Spectrofluorometer at $25°C$ with fluorescent probe concentrations ranging between 0.25 and $2.5\mu M$[31]. S-1 and actin concentrations were between 10 and $15\mu M$. The Stern-Volmer quenching constants, K_{sv}, were determined according to the quenching expression

$$F/Fo = 1 + K_{sv} [Q]$$

where Fo is the fluorescence intensity in the absence of quencher and F is the fluorescence in the presence of a given concentration of a quencher [Q].

Fluorescence measurements in the presence of anti-dansyl (anti-DAZ) and anti-fluorescein (anti-Fl) antibodies were carried out as described above except for the absence of acrylamide and nitromethane. Small aliquots of anti-DAZ and anti-Fl were added to the appropriate samples. All fluorescence intensity readings were corrected for dilution effects.

Binding Experiments

Solutions of DAZ-F-actin and Fl-F-actin in 30mM KCl, 2mM $MgCl_2$, 10mM Bis-Tris pH 7.2 were incubated for 30 min. at $25°C$ with anti-DAZ and anti-Fl IgG in the presence and absence of S-1. Actin concentrations were between 4.0 and $12\mu M$ and S-1, when present, was equimolar with actin. The antibodies were equimolar with the concentrations of the labeled actin (between 50 and 70% of total actin concentration). All samples were centrifuged at 140,000g for 20 min. in a Beckman airfuge. The solubilized pellets and supernatant fractions were denatured and run on 10% SDS polyacrylamide gels[32]. The molar ratios of S-1, IgG, and F_{ab} (1-7) bound to actin were obtained from densitometric analysis of the Coomassie-Blue stained protein bands[13].

The competition between S-1 and anti-Fl IgG for the binding to Fl-actin was tested also in competitive ELISA experiments[11]. The wells of Dynatech Immulon microplates were coated with $5\mu g$ of Fl-F-actin. The amounts of S-1 added in the presence of IgG ranged between 0 and $1,200\mu g$.

Measurements of Actin-Activated ATPase Activity of S-1

Acto-S-1 ATPase activities were measured at $25°C$ in G-actin buffer, in the presence of 2mM MgATP, as previously described[13]. Fl-actin and S-1 concentrations were 4.5 μM, F_{ab} (1-7) was $13\mu M$ and anti-Fl IgG, when present, was set at $3.0\ \mu M$ ($30\mu M$ total IgG concentration).

RESULTS

Accessibility of AEDANS and N-Dansylaziridine on Cys-374 in Actin

Steady-state quenching experiments were carried out in order to determine the accessibility in actin and acto-S-1 of fluorescent probes attached to Cys-374 on actin. As shown in Figure 1, acrylamide quenched readily the fluorescence intensity of free IAEDANS ($K_{sv} = 10.0 \pm 0.2M^{-1}$). The quenching was reduced about 3-fold ($K_{sv} = 3.3 \pm 0.1M^{-1}$) when the probe was attached to G-actin. It was reduced further upon polymerization of the labeled actin into filaments ($K_{sv} - 1.8 \pm 0.1M^{-1}$). The binding of S-1 to F-actin did not produce any additional change in the accessibility of the AEDANS probe to acrylamide (Figure 1). We have verified by airfuge pelleting

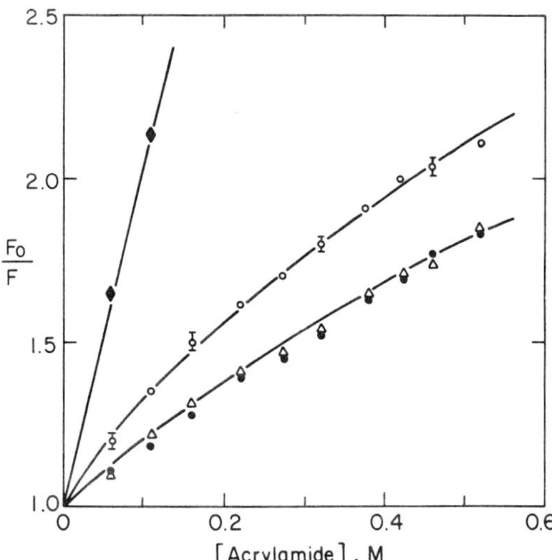

Figure 1: Stern-Volmer plots for the quenching of actin-AEDANS fluorescence by acrylamide. Free IAEDANS (♦), AEDANS-G-actin (○), AEDANS-F-actin (●), and AEDANS-acto-S-1 (▵) were titrated with acrylamide in G-actin buffer at 25°C. F-actin and acto-S-1 solutions contained also 2mM $MgCl_2$. Each sample contained 0.5μM AEDANS moiety. The concentrations of actin and S-1 were 11.0μM each. The excitation and emission wavelengths were 337 and 480 nm respectively.

experiments that the binding of S-1 to ADEANS-F-actin and the stability of actin filaments were not affected in a significant manner by up to 0.5M acrylamide[30]. Also, in agreement with the previous study[30], a downward curvature was detected in the quenching titrations of actin indicating the presence of multiple emission components. The analysis of these components[30] revealed that the initial slope of an intensity quenching plot for AEDANS-actin was to a good approximation equal to K_{sv} of the major decay component.

The second probe, N-dansylaziridine, which also modifies Cys-374 on actin[29], has a similar two-ring structure but is less polar that IAEDANS. Nevertheless, as shown in Figure 2, the quenching of DAZ-actin by nitromethane was similar to that observed for AEDANS-actin (Figure 1). K_{sv} values for free DAZ-cysteine and DAZ-G-actin were 18 ± 0.3 and $6.3 \pm 0.2M^{-1}$, respectively. The dansyl probes on F-actin and in acto-S-1 showed equally reduced accessiblity to the quencher ($K_{sv} = 3.6 \pm 0.1M^{-1}$). As monitored by airfuge pelleting experiments, nitromethane (up to 0.4 M), did not have any significant effect on acto-S-1 binding and the stability of actin filaments (data not shown).

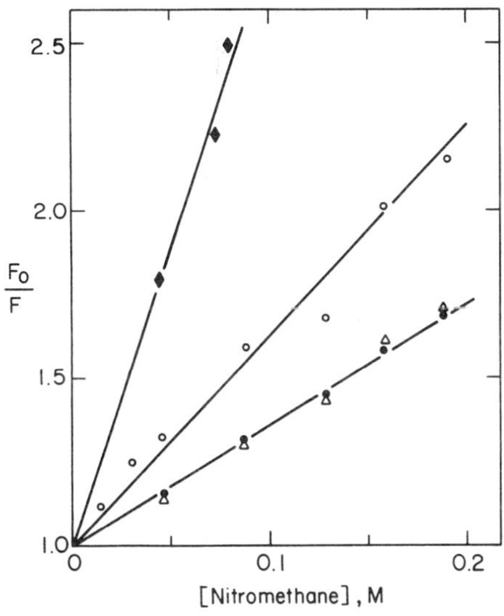

Figure 2: Stern-Volmer plots for the quenching of DAZ-actin fluorescence by nitromethane. DAZ-cysteine (♦), DAZ-G-actin (○), DAZ-F-actin (●), and DAZ-acto-S-1 (△), were titrated with nitromethane in G-actin buffer. Each sample contained 2.5 μM DAZ moiety. The concentrations of actin and S-1 were 12.5 μM each. λ_{Ex} and λ_{Em} were 350 and 495 respectively.

The environment of the dansyl probe on actin was tested also with monoclonal anti-dansyl antibodies. These experiments were prompted by the previous observations that the binding of anti-dansyl IgG to the probe enhances its fluorescence[33]. As shown in Figure 3, the fluorescence of DAZ-G-actin increased by about 45% upon addition of anti-DAZ IgG. In contrast to this, the same antibodies did not elicit any changes in the fluorescence of DAZ-F-actin and DAZ-acto-S-1. For the free DAZ-cysteine complex the fluorescence increase due to antibody binding was about 10-fold. Anti-DAZ IgG had almost no effect on the fluorescence of AEDANS-G-actin or the free IAEDANS.

The fluorescence titrations of DAZ-actin with anti-DAZ IgG suggested the inaccessibility of the dansyl probe on F-actin and in acto-S-1 to the antibodies. This was confirmed by airfuge centrifugation experiments. The pelleting of mixtures of DAZ-F-actin and DAZ-acto-S-1 with anti-DAZ IgG did not reveal any cosedimentation of antibodies with actin. Thus, the dansylaziridine moiety on F-actin and in acto-S-1 was not available for antibody binding.

Interactions of Fluorescein-labeled Actin with S-1

The modification of Cys-374 with iodoacetamidofluorescein introduced a more hydrophilic probe into the C-terminus region of actin than the two dansyl-based reagents reported above. This probe was expected to be less buried in the interior of the protein than the dansyl reagents. Quenching titrations of Fl-actin with

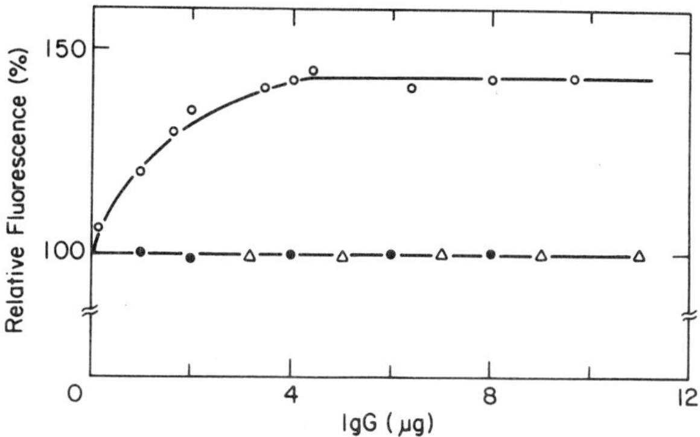

Figure 3: Enhancement of fluorescence intensity of DAZ-actin by monoclonal anti-DAZ IgG. The concentration of actin in G-actin (○), F-actin, (●), and acto-S-1 (▲) solutions was 11.0 μM. S-1 was set also at 11.0μM. λ_{Ex} = 350nm; λ_{Em} = 495 nm.

nitromethane (Figure 4) confirmed this expectation. The quenching constants determined for a free reagent (K_{sv} = 7.8 ± 0.2M^{-1}), Fl-G-actin (K_{sv} = 4.5 ± 0.1M^{-1}), and Fl-F-actin (K_{sv} = 3.4 ± 0.1M^{-1}) were significantly closer to each other than in the case of AEDANS-and DAZ-actin. This indicated that the change in the environment of the Cys-374 probe upon the polymerization of actin depended to a large extent on the probe itself. Additional evidence for this point was provided by examining the fluorescence intensity ratios F_f/F_g for F-actin and G-actin. The F_f/F_g ratios were 2.10 ±0.18, 1.40 ± 0.20, and 1.08 ± 0.04 for AEDANS-, DAZ-, and Fl-actin, respectively, indicating a decreasing perturbation, in that order, of the probe region on polymerization of actin. Also, in contrast to AEDANS- and DAZ-actin, S-1 caused a small but reproducible additional protection of Fl-F-actin from quenching by nitromethane (K_{sv} = 2.9 ± 0.1 M^{-1}; Figure 4).

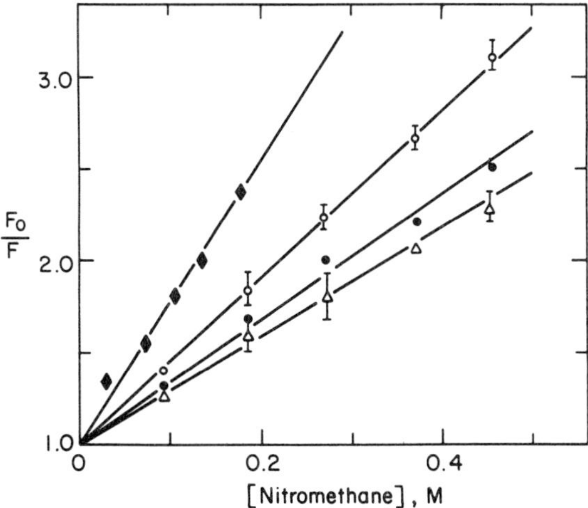

Figure 4: Stern-Volmer plots for the quenching of Fl-actin by nitromethane. Free iodoacetamide fluorescein (♦), Fl-G-actin (○), Fl-F-actin (●), and Fl-acto-S-1 (△) were titrated with nitromethane in G-actin buffer. The concentration of Fl-moiety was 0.5μM and actin and S-1 were set at 11.0 μM each. λ_{Ex} = 365 nm; λ_{Em} = 517 nm.

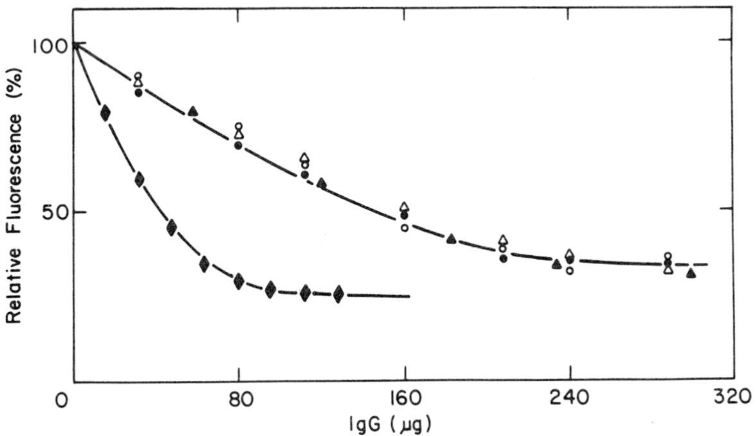

Figure 5: Quenching of the fluorescence of Fl-actin by anti-Fl IgG. Free iodoacetamide fluorescein (♦), Fl-G-actin (○), Fl-F-actin (●), Fl-F-actin-F_{ab}(1-7) (△), and Fl-acto-S-1 (▲) were titrated with IgG containing 12% anti-Fl specific antibodies. The concentrations of actin, S-1, and F_{ab}(1-7) were 5.0, 50, and 15.0 μM respectively. λ_{ex} = 365 nm; λ_{em} = 517 nm.

The effect of S-1 on the fluorescein probe on actin was assessed further by using anti-fluorescein antibodies. The binding of specific antibodies to fluorescein was shown before to quench the fluorescence signal[27]. Thus, with the decrease in fluorescence intensity as an indicator of antibody binding, the accessibility of fluorescein on Cys-374 to anti-Fl IgG could be readily measured. As shown in Figure 5, the binding of anti-Fl IgG to fluorescein quenched its fluorescence by about 70%. The titration curves revealed that the antibody had a higher affinity for the free iodoacetofluorescein than for the actin-attached probe. More importantly, the antibodies appeared to bind equally well to fluorescein on G- and F-actin and even on acto-S-1. Clearly, neither the polymerization of actin nor the presence of a 10-fold (Figure 5) or 20-fold (not shown) molar excess of S-1 over actin changed significantly the binding surface presented to antibodies by the probe on Cys-374.

The lack of any S-1 effect on anti-Fl binding to F-actin, called for a direct documentation of S-1 binding to the actin-anti-Fl IgG complex. This was done by airfuge centrifugation of mixtures of F-actin (4.0μM), anti-Fl IgG (1.9μM), and S-1 (between 5 and 80 μM). The pelleted samples were examined on SDS polyacrylamide gels and clearly showed that S-1 was sedimented together with actin and IgG. Densitometric analysis of these gels yielded the binding information summarized in Table 1.

Table 1: Binding of S-1, F_{ab}(1-7), and Anti-Fl IgG to F-actin

Protein Added (μM)	Protein Present (μM)	Protein Bound to Actin[a]		
		S-1	Fab(1-7)	IgG
S-1 (5-80μM)	IgG (1.9μM)	1.01 ± 0.1		0.35 ± 0.03
---	IgG (1.9μM)			0.37 ± 0.04
Fab(1-7) (12μM)	IgG (1.9μM)		0.72 ± 0.05	0.38 ± 0.03

[a] Molar ratios of S-1, F_{ab}(1-7), and anti-Fl IgG bound to F-actin in 30mM KCl, 2mM $MgCl_2$, 10mM Bis-Tris (pH 7.2) were determined as described in Materials and Methods. The concentration of Fl-actin was 4.0μM; concentrations of other proteins are given above.

Strikingly, over a wide range of concentrations, S-1 did not have any significant effect on the binding of anti-Fl IgG to the labeled actin. The binding of antibodies to actin (0.35 IgG/actin), in turn, did not decrease the binding of S-1 to actin. These results were consistent with fluorescence titrations of actin with IgG (Figure 5) and indicated that under rigor conditions the myosin binding site on actin does not overlap with the fluorescein probe on Cys-374.

ELISA titrations of Fl-actin with anti-Fl IgG in the presence of S-1 (up to 1.2 mg/ml) revealed an inhibition by S-1 of antibody binding to the adsorbed actin. This inhibition was similar, to that observed for antidansyl antibodies[9], suggesting that the local conformations of Cys-374 environment in solid phase and solution might be different.

In order to test for a possible effect of anti-Fl antibodies on weak actomyosin interactions in the presence of MgATP, the actin activated ATPase activity was measured in the presence and absence of anti-Fl IgG. Up to 0.48 molar ratios of IgG bound to Fl-actin, which for a divalent antibody could correspond to even higher fractional saturation of fluorescein sites on actin, did not inhibit the acto-S-1 ATPase activity (Table 2). Thus, although in general the weak acto-S-1 interactions are perturbed more readily than rigor actomyosin bonds[13,11,34,35], the anti-Fl antibody did not have any effect on the weak acto-S-1 binding.

Table 2: Effect of Antibodies on Acto-S-1 ATPase Activity

Proteins	V^a (s^{-1})
Actin, S-1	2.6
Actin, S-1, ant-Fl IgG	2.8
Actin, S-1, Fab(1-7)	0.5
Actin, S-1, Fab(1-7), anti-Fl IgG	0.6

[a]Actin-activated MgATPase activities of S-1(V) were determined in G-actin buffer, at 25°C, in the presence of 2mM MgATP. The turn-over rates reported above were corrected for the ATPase activity of S-1 alone. Actin and S-1 concentrations were 4.5μM; Fab(1-7) and anti-Fl IgG were added to final concentrations of 13.0 and 3.0μM respectively. The molar ratios of Fab(1-7) and IgG bound to actin were 0.75 and 0.48 respectively.

Fluorescein-labelled C-terminus of Actin and its N-terminal Region

The proximity of the C- and N-termini of actin raises a possibility of signal transmission between these two adjacent regions on the protein. Since the N-terminus of actin contains functionally important acidic residues[35], its indirect perturbation (via the C-terminus) could modulate macromolecular interactions of actin. In order to test for structural coupling between the N- and C-termini of actin, the binding of anti-Fl IgG to Fl-actin was measured in the presence and absence of antibodies against residues 1-7 on actin, $F_{ab}(1-7)$. Approximately 70-75% saturation of actin by $F_{ab}(1-7)$ did not have any effect on the binding of anti-Fl IgG to actin (Table 1) and on the quenching of the fluorescein fluorescence by anti-Fl IgG (Figure 5). Conversely, IgG did not displace F_{ab}(1-7) from actin as judged by the lack of any modulation of F_{ab}(1-7) effect on weak acto-S-1 interactions in the presence of MgATP. F_{ab}(1-7) inhibited the acto-S-1 ATPase activity to the same extent (by about 80%) in the absence and presence of 0.48 mol of IgG bound per mol of actin (Table 2). These results show, using fluorescein attached to Cys-374 as a probe, that N- and C-termini of actin are not coupled structurally.

DISCUSSION

Several lines of evidence suggested that the C-terminus of actin is located either close to or directly at the myosin binding interface. Among experiments supporting such a conclusion are the cross-linking of Cys-374 on G-actin to S-1[21], inhibition of antibody binding to AEDANS-actin by S-1[9], and the well documented quenching of pyrenyl-actin fluorescence by S-1[36]. Yet, as in most cases, it is difficult to discern whether indirect effects or the binding of S-1 to the C-terminus of actin are responsible for these and other observations. The former possibility is indicated by the fact that tryptic removal of 2 or 3 C-terminal residues does not change much acto-S-1 interactions[22-24]. The actual role of Cys-374 in macromolecular interactions of actin is of particular interest since the atomic structure of actin has been solved for a protein with the last 3-residues removed by carboxypeptidase[5].

By using three different fluorescent probes targeted at Cys-374 we have shown that the environment of this residue and the changes in it are probe-dominated. Dansylaziridine is "buried" in the F-actin structure and is not available for antibody binding in the presence and absence of S-1. Thus, the 20% increase in DAZ-F-actin fluorescence observed on S-1 binding reflects most likely an indirect change in the Cys-374 region propagated from the acto-S-1 interface. In contrast to dansylaziridine the fluorescein probe is equally accessible to antibodies in G-actin, F-actin, and acto-S-1. Nevertheless, some differences in the environment of fluorescein on residue Cys-374 in the three states of actin are indicated by the results of collisional quenching titrations. Since these differences are too small to affect the affinities of anti-Fl IgG for the probe, the energy barriers between conformational states of fluorescein-modified C-terminus in G-actin, F-actin, and acto-S-1 must be rather small.

If the spectral perturbations of Cys-374 probes by S-1 are due mostly to signal transmission from other regions of the protein, the adjacent N-terminus of actin, which binds to myosin[35], would appear to be a good candidate for a conformational coupling with the C-terminus of actin. Such a coupling has not been detected by using F_{ab}(1-7) and anti-Fl IgG neither in the presence nor the absence of MgATP. Thus, the changes in actin which result from its binding to myosin appear to be transmitted to the C-terminal region from sites other than the N-terminus of actin. These sites may include sequences 21-29 and 91-103[9,16] but probably not the proximal residues 361-364 on actin[16]. The recently demonstrated signal transmission from the divalent cation site on actin to its C-terminus[37] provides an example for such a conformational coupling between different regions on actin. The differences in the properties of actin modified with different probes and the implicit flexibility of actin's C-terminus merit special attention in designing probe-dependent experiments on actin's structure and dymanics.

ACKNOWLEDGMENTS

This work was supported by USPHS grant AR 22031 and NSF grant MCB 9206739.

REFERENCES

1. R.K. Cook, W.T. Blake, and P.A. Rubenstein, Removal of the amino-terminal acidic residues of yeast actin, J. Biol. Chem., 267:9430 (1992).
2. P. Aspenstrom, and R. Karlsson, Interference with myosin subfragment-1 binding by site-directed mutagenesis of actin, Eur. J. Biochem., 200:35 (1992).
3. K. Sutoh, M. Ando, K. Sutoh, and Y.Y. Toyoshima, Site-directed mutations in Dictyostelium actin: disruption of a negative charge cluster at the N-terminus, Proc. Natl. Acad. Sci. USA, 88:7711 (1991).
4. K.F. Wertman, D.G. Drubin, and D. Botstein, Systematic mutational analysis of the yeast ACT 1 gene, Genetics, 132:337 (1992).
5. W. Kabsch, H.G. Mannherz, D. Suck, E. Pai, and K.C. Holmes, Atomic structure of the actin-DNase I complex, Nature, 347:37 (1990).
6. W. Kabsch, and J. Vandekerchove, Structure and function of actin, Annu. Rev. Biophys. Biomol. Struct., 21:49 (1992).
7. R. Bertrand, P. Chaussepied, E. Audemard, and R. Kassab, Functional characterization of skeletal F-actin labeled on the NH_2-terminal segment of residues 1-28, Eur. J. Biochem., 181:747 (1989).
8. J.E. Van Eyk, and R. Hodges, A synthetic peptide of the N-terminus of actin interacts with myosin, Biochemistry, 30:11676 (1991).
9. C. Mejean, M. Boyer, J.P. Labbe, J. Derancourt, Y. Benyamin, and C. Roustan, Anti-actin antibodies: an immunological approach to the myosin-actin and tropomyosin-actin interfaces, Biochem. J., 244:571 (1987).
10. J.P. Labbe, C. Mejean, Y. Benyamin, and C. Roustan, Characterization of an actin-myosin head interface in the 40-113 region of actin using specific antibodies as probes Biochem. J., 271:407 (1990).
11. L. Miller, M. Kalnoski, Z. Yunossi, J.C. Bulinski, and E. Reisler, Antibodies directed against N-terminal residues on actin do not block acto-myosin binding, Biochemistry, 26:6064 (1987).
12. G. DasGupta, and E. Reisler, Antibody against the amino terminus of α-actin inhibits actomyosin interactions in the presence of ATP, J. Mol. Biol., 207:833 (1989).
13. G. DasGupta, and E. Reisler, Acto-myosin interactions in the presence of ATP and the N-terminal segment of actin, Biochemistry, 31:1836 (1992).
14. A.J.G. Moir, and B.A. Levine, Protein cognitive sites on the surface of actin. A proton NMR study, J. Inorganic Chem., 27:271 (1986).
15. R.K. Cook, D. Root, C. Miller, E. Reisler, and P.A. Rubenstein, Enhanced stimulation of myosin subfragment 1 ATPase activity by addition of negatively charged residues to the yeast actin NH_2 terminus, J. Biol. Chem., 268:2410 (1993).
16. M. Johara, Y.Y. Toyoshima, A. Ishijima, H. Kojima, T. Yanagida, and K. Sutoh, Charge-reversion mutagenesis of Dictyostelium actin to map the surface recognized by myosin during ATP-driven sliding motion, Proc. Natl. Acad. Sci. USA, 90:2127 (1993).
17. T. Katoh, and S. Lowey, Mapping myosin light chains by immunoelectron microscopy. Use of anti-fluorescyl antibodies as structural probes, J. Cell Biol., 109:1549 (1989).
18. B. Malm, L.E. Nystrom, and U. Lindberg, The effect of proteolysis on the stability of the profilactin complex, FEBS Lett., 113:241 (1980).
19. P. Graceffa, and A. Jancso, Disulfide cross-linking of caldesmon to actin, J. Biol. Chem., 266:20305 (1991).
20. Y. Doi, M. Banba, and A. Vertut-Doi, Cysteine 374 of actin resides at the gelsolin contact site in the EGTA resistant actin-gelsolin complex, Biochemistry, 30:5769 (1991).
21. C. Combeau, D. Didry, and M.-F. Carlier, Interaction between G-actin and myosin subfragment-1 probed by covalent cross-linking, J. Biol. Chem., 267:14038 (1992).
22. R.H. Crosbie, J.M. Chalovich, and E. Reisler, Interaction of caldesmon and myosin subfragment 1 with the C-terminus of actin, Biochem. Biophys. Res. Commun., 184:239 (1992).
23. R. Makuch, J. Kotakowski, and R. Dabrowska, The importance of C-terminal amino acid residues of actin to the inhibition of actomyosin ATPase activity by caldesmon and troponin I, FEBS Lett., 297:237 (1992).
24. S.I. O'Donoghue, M. Miki, and C.G. DosRemidios, Removing the two C-terminal residues of actin affects the filament structure, Arch. Biochem. Biophys. 293:110 (1992).
25. A.G. Weeds, and B. Pope, Studies on the chymotryptic digestion of myosin. Effects of divalent cations on proteolytic susceptibility, J. Mol. Biol., 111:129 (1977).
26. J.A. Spudich, and S. Watt, Regulation of skeletal muscle contraction. I. Biochemical studies of the interaction of the tropomyosin-troponin complex with actin and the proteolytic fragments of myosin, J. Biol. Chem., 246:4866 (1971).

27. D.E. Lopatin, and E.W. Voss, Jr., Fluorescein. Hapten and antibody active-site probe, Biochemistry, 10:208 (1971).
28. R. Takahashi, Fluorescence energy transfer between subfragment-1 and actin points in the rigor complex of acto subfragment-1, Biochemistry, 18:5164 (1979).
29. T.-I. Lin, Fluorimetric studies of actin-labeled with dansyl aziridine, Arch. Biochem. Biophys., 185:285 (1978).
30. T. Tao, and J. Cho, Fluorescence lifetime quenching studies on the accessibilities of actin sulfhydryl sites, Biochemistry, 18:2759 (1979).
31. L. Miller, M. Phillips, and E. Reisler, Polymerization of G-actin by myosin subfragment 1, J. Biol. Chem., 263:1996 (1988).
32. U.K. Laemmli, Cleavage of structural proteins during the assembly of the head of bacteriophage T4, Nature (London) 227:680 (1970).
33. J. Reidler, V.T. Oi, W. Carlsen, T.M. Vuong, I. Pecht, L.A. Herzenberg, and L. Stryer, Rotational dynamics of monoclonal anti-dansyl immunoglobulins, J. Mol. Biol., 158:739 (1982).
34. G. DasGupta, and E. Reisler, Nucleotide-induced changes in the interaction of myosin subfragment 1 with actin: Detection by antibodies against the N-terminal segment of actin, Biochemistry, 30:9961 (1991).
35. E. Reisler, Actin molecular structure and function, Curr. Opin. Cell Biol., 5:41 (1993).
36. T. Kouyama, and K. Mihashi, Fluorimetry study of N-(1-pyrenyl)iodoacetamide-labeled F-actin, Eur. J. Biochem. 114:33 (1981).
37. M. Mossakowska, J. Moraczewska, S. Khaitlina, H. Strzelecka-Golaszewska, Proteolytic removal of three C-terminal residues of actin alters the monomer-monomer interactions, Biochem. J. 289:897 (1992).

ACTIN POLYMERIZATION: REGULATION BY DIVALENT METAL ION AND NUCLEOTIDE BINDING, ATP HYDROLYSIS AND BINDING OF MYOSIN

Marie-France Carlier, Catherine Valentin-Ranc, Cecile Combeau
Stephane Fievez and Dominique Pantoloni

Laboratoire d'Enzymologoie, C.N.R.S.
91198 Gif-sur-Yvette Cedex, France

SUMMARY

Actin filaments are major dynamic components of the cytoskeleton of eukaryotic cells. Assembly of filaments from monomeric actin occurs with expenditure of energy, the tightly bound ATP being irreversibly hydrolyzed during polymerization. This dissipation of energy perturbs the laws of reversible helical polymerization defined by Oosawa and Asakura (1975), and affects the dynamics of actin filaments. We have shown that ATP hydrolysis destabilizes actin-actin interactions in the filament. The destabilization is linked to the liberation of P_i that follows cleavage of the γ–phosphate. P_i release therefore plays the role of a conformational switch. Because ATP hydrolysis is uncoupled from polymerization, the nucleotide content of the filaments changes during the polymerization process, and filaments grow with a stabilizing "cap" of terminal ADP-P_i subunits. The fact that the dynamic properties of F-actin are affected by ATP hydrolysis results in a non-linear dependence of the rate of filament elongation on monomer concentration.

Possible modes of regulation of filament assembly may be anticipated from the basic properties of actin. We have shown that the tightly bound divalent metal ion (Ca^{2+} or Mg^{2+}) interacts with the β- and γ-phosphates of ATP bound to actin, and that the Me-ATP bidentate chelate is bound to G-actin in the Λ configuration. The nature of the bound metal ion affects the conformation of actin and the rate of ATP hydrolysis.

In motile living cells, a large pool of actin is maintained unpolymerized by interaction with G-actin binding proteins such as thymosin $\beta 4$ and its variants or profilin. Part of this pool is released to increase the F-actin pool upon cell stimulation. The role of G-actin polymerizing proteins may be crucial in defining the patterns of filament assembly in these situations. The myosin head (myosin subfragment-1) may be

Actin: Biophysics, Biochemistry, and Cell Biology
Edited by J.E. Estes and P.J. Higgins, Plenum Press, New York, 1994

considered as a model actin polymerizing protein, may be the closest model to the short tailed myosin I family. The mechanism of assembly of decorated filaments from G-actin and myosin subfragment-1 has therefore been examined.

INTRODUCTION

In eukaryotes, cellular space is organized by an ensemble of fibrous polymers called cytoskeleton. Actin filaments and microtubules, which are major components of the cytoskeleton, are dynamic structures, in contrast to most structural macromolecular assemblies, such as viruses. Their assembly and disassembly processes account for many motile activities of the cell. In addition, these helical polymers are dissipative structures: assembly of the polymer is coupled to energy consumption, in the form of ATP or GTP hydrolysis.

Monomeric actin binds ATP very tightly ($K_A = 10^{10}$ M^{-1} in low ionic strength buffers, in the presence of Ca^{2+} ions). A polymerization cycle involves addition of the ATP-monomer to the polymer end, hydrolysis of ATP on the incorporated subunit, liberation of P_i in solution, and dissociation of the ADP-monomer. Exchange of ATP for bound ADP occurs on the monomer only, and precedes its being committed to another polymerization cycle. Therefore monomer-polymer exchange reactions are performed with expenditure of energy, exactly one mol of ATP per mol of actin incoporated in actin filaments. As a result, up to 40 % of the ATP consumed in motile cells is used to maintain the dynamic state of actin. It is therefore important to understand how the free energy of nucleotide hydrolysis is utilized in cytoskeleton assembly.

ATP HYDROLYSIS ASSOCIATED TO ACTIN POLYMERIZATION PERTURBS THE THERMODYNAMICS OF REVERSIBLE POLYMERIZATION

a) ATP hydrolysis is linked to the destabilization of actin filaments and microtubules.

The critical concentration, that is the monomer $\Leftrightarrow$ polymer equilibrium dissociation constant for polymerization of ADP-actin is 25-fold larger than for polymerization of ATP-actin. However, in both cases the filament is made of F-ADP subunits, and the rate constant for association of ADP-actin to filament ends is only 2.5 fold lower than the rate constant for association of ATP-actin. In the absence of free ATP, the 1:1 ATP-actin complex can polymerize, but the polymer once formed spontaneously depolymerizes. Depolymerization stops when the concentration of ADP-monomer in the medium reaches the value of the critical concentration for polymerization of ADP-actin (for review see Korn *et al.*, 1987; Carlier, 1991).

The above observations are inconsistent with a simple two-state polymerization model within which only two species, ATP-G-actin and ADP-F-actin, coexist in solution.

b)   Thermodynamic and kinetic parameters for reversible polymerization (Oosawa's law)

The theory of reversible helical polymerization of proteins has been fully described by Oosawa (1975). The following equation describes polymer growth :

$$J(c) = dc/dt = k_+ [P] c - k_- [P] \qquad (1)$$

where $J(c)$ is the rate of polymer growth, c and P represent the concentrations of monomer and polymer elongating sites respectively: k_+ and k_- are the rate constants for monomer association to and dissociation from polymer ends. According to equation (1), k_+ and k_- can easily be derived from the linear dependence of $J(c)$ on c, and the critical concentration $c_c = k_- / k_+$ defined as the monomer concentration at which $J(c) = 0$.

c)   Thermodynamic and kinetic parameters for ATP-actin polymerization

The polymer growth $J(c)$, showed nonlinear monomer concentration dependence in the presence of ATP (Carlier *et al.*, 1984), while in the presence of ADP, the plot of $J(c)$ versus monomer concentration for actin was a straight line, as expected for a reversible polymerization. The data imply that newly incorporated subunits dissociate from the filament at a slower rate than internal ADP-subunits: in other words (1) the effect of nucleotide hydrolysis is to decrease the stability of the polymer by increasing k_-; (ii) nucleotide hydrolysis is uncoupled from polymerization and occurs in a step that follows incorporation of a ATP-subunit in the polymer. Newly incorporated, slowly dissociating, terminal ATP-subunits form a stable "cap" at the ends of F-actin filaments.

The above results demonstrate that ATP hydrolysis associated with actin and tubulin polymerization acts as a regulatory switch affecting the strength of protein-protein interactions. In this respect, this biological system appears similar to the G-proteins or other regulatory nucleoside triphosphatases. Indeed, one can consider that actin exists in two states: a "non-interacting" state, in which ATP-G-actin does not hydrolyse nucleotide, and an "interacting" state in which F-actin hydrolyses ATP in a single turnover reaction, and nucleotide hydrolysis is linked to a weakening of actin-actin interactions in the polymer lattice.

In order to anticipate possible modes of regulation of cytoskeleton dynamics *in vivo*, it is necessary: (i) to identify the kinetic intermediates involved in the polymerization process and to characterize their structural and functional properties; (ii) to define the essential elementary steps in the hydrolysis process.

d)   Kinetic steps in ATP hydrolysis on F-actin

ATP is hydrolysed in at least two consecutive steps on F-actin, cleavage of the γ-phosphoester bond, followed by P_i release, according to the following scheme :

$$F\text{-}ATP \Leftrightarrow F\text{-}ADP\text{-}P_i \Leftrightarrow F\text{-}ADP + P_i \qquad (2)$$

P_i release occurs at a relatively apparent slow rate ($k_{obs} = 0.005$ s^{-1}), so that the transient intermediate F-ADP-P$_i$, in which P$_i$ is non covalently bound, has a life time of 2-3 min (Carlier and Pantaloni, 1986; Carlier, 1987). While the γ-phosphate cleavage step is irreversible as assessed by ^{18}0 exchange studies (Carlier *et al.*, 1987), the release of P$_i$ is reversible. Binding of H_2PO_4 (K_p 10^{-3}M) causes the stabilization of actin filaments and the rate of filament growth varies linearly with the concentration of actin monomer in the presence of P$_i$ (Carlier & Pantaloni, 1988). Therefore P$_i$ release appears as the elementary step responsible for the destabilization of actin-actin interactions in the filament.

e)   Probing the intermediate ADP-P state on F-actin using structural analogues of P_i: AlF_4^- and BeF_3^-, H_2O

Fluoroaluminate and fluoroberyllate have a tetrahedral configuration in solution with the bond lengths similar to those of inorganic phosphate, and have been shown to restore the functional properties of GTP-transducin when added to GDP-transducin (Bigay *et al.* , 1987). These phosphate analogues bind to F-ADP-actin in competition with P_i, but with an affinity three orders of magnitude higher than P_i (Combeau and Carlier, 1988, 1989). The F-ADP-BeF$_3$ filaments are extremely stable, the rate of dissociation of ADP-BeF$_3$ subunits from filament ends is very low, actually even lower than the rate of dissociation of ADP-P$_i$ subunits. In addition, some evidence suggests that the conformation of the F-ADP-BeF$_3^-$ state is different from that of the F-ADP-P$_i$ state. BeF$_3^-$ and AlF$_4^-$ bind to and dissociate from the ADP-polymer at very slow rates. All the above properties of BeF$_3^-$ and AlF$_4^-$ are very similar to those of vanadate in other ATPases, e.g. vanadate binding to ADP-myosin (Goodno 1979) and led to the suggestion that BeF$_3^-$ and AlF$_4^-$ could mimic the ADP-P* transition state, or at least adopt a configuration closer to that of bound ATP than to bound ADP-P$_i$. Similar results have been obtained for binding of BeF$_3^-$ and AlF$_4^-$ to the bacterial F$_1$-ATPase (Dupuis *et al.*, 1989) and to myosin (Phan and Reisler, 1992). Further experiments should be aimed at understanding the structure of bound ADP-AlF$_4^-$, for example using the superhyperfine coupling of Mn ESR signal with ^{17}O labeled ADP, or using NMR of ^{19}F. Interestingly, AlF$_4^-$ and BeF$_3^-$ do not bind to monomeric G-ADP-actin which is not able to hydrolyse the nucleotide. This observation indicates that the environment of the γ-phosphoester bond of the nucleotide is not the same in the monomer and in the polymerized states of actin.

The results of the experiments using phosphate analogues lead to add another step in the kinetic scheme for hydrolysis of ATP or GTP on F-actin or microtubules.

$$F\text{-}ATP \rightarrow F\text{-}ADP\text{-}P^* \Leftrightarrow F\text{-}ADP\text{-}P_i \Leftrightarrow F\text{-}ADP\text{-}P_i \tag{3}$$

In the above scheme, F-ADP-P* represents the transition state energetically identical to the F-ADP BeF$_3^-$ state. The transition from F-ADP-P* to F-ADP-P$_i$ would be slow and rate limiting for P$_i$ release. In this scheme, which resembles the one proposed for ATP hydrolysis on myosin for example (Hibberd & Trentham 1986), P$_i$ binds to F-ADP in rapid equilibrium, while dissociation of P$_i$ following cleavage of ATP is slow.

f)   Mechanistic models for ATP hydrolysis in F-actin assembly

ATP may potentially be hydrolyzed in several ways following the incorporation of an ATP-actin subunit in the filament : the rate of ATP hydrolysis may be independent of the nature of the nucleotide bound to neighboring subunits (ATP or ADP), which can be called "random hydrolysis", or it may be affected, by "induced-fit", by the conformation of neighboring subunits, i.e. by the bound nucleotide. An extreme case is the one where hydrolysis occurs at a much faster rate on an ATP-subunit distally adjacent to an ADP-subunit : in this "vectorial" hydrolysis model, ATP hydrolysis occurs essentially at the ATP cap/ADP core boundary migrates distally like a zipper at a constant rate. This latter theoretical model appears to adequately account for the data obtained with MgATP-actin (Carlier *et al.*, 1986, 1987) with the additional formation of new ATP/ADP boundaries taking place at high rate of filament growth, via random hydrolysis in long stretches of rapidly assembled F-ATP-actin. In contrast, the data obtained with Ca-ATP-actin are essentially described by a model of random hydrolysis of ATP on any F-ATP subunit independently of the nature of the neighboring subunit; hence the cap of F-ATP is larger on filaments growing from Ca-ATP-actin than on filaments growing from MgATP-actin, other medium conditions being the same (0.1 M KCl).

g)   Structural change of F-actin associated to ATP hydrolysis

The results from thermodynamic and kinetic studies on actin and tubulin polymerization indicate that a structural change of the polymer is linked to P_i release. The nature of this change is a challenging issue. In a recent study combining electron microscopy and image reconstruction from negatively stained F-ADP and F-ADP-BeF_3^- filaments, a structural change localized in subdomain 2 of the actin subunit has been detected (Orlova and Egelman, 1992). Evidence for different structural states of the filament in the F-ADP, F-ADP-P_i and F-ADP-BeF_3^- states can also be obtained by a combination of cryoelectromicroscopy and solution low angle X-ray scattering techniques (Lepault *et al.* submitted). This structural change is expected to be less spectacular than in the case of ras p21, because the conformation of the subunit is somewhat constrained in the NTP state by the structure of the polymer itself. The change in the coordination of the divalent metal ion following the release of P_i is likely to trigger this structural change.

In the three-dimensional structure of actin, the environment of the phosphate moiety of the nucleotide appears roughly the same whether CaADP or CaATP is bound, whereas one would expect to observe two different conformations. The reason for this finding is unclear, however it must be stressed that the three-dimensional structure is derived from X-ray diffraction of crystals of the DNaseI-actin complex, which is, like G-actin, unable to hydrolyse ATP. The conformation obtained may therefore correspond to G-actin frozen in the G-ATP state independently of the bound nucleotide. Structural studies in conjunction with site-directed mutagenesis experiments should eventually solve the issue.

THE CRITICAL CONCENTRATION INCREASES WITH THE NUMBER OF FILAMENTS IN THE PRESENCE OF ATP

In reversible polymerization the critical concentration is equal to the equilibrium dissociation constant for polymer formation. This thermodynamic parameter is therefore independent of the number of polymers in solution. This law is unquestionably verified for the reversible polymerization of ADP-actin: when sonic vibration is applied to a solution of F-ADP-actin filaments at equilibrium with G-ADP monomers, no change is observed in the proportion of G- and F-actin (Carlier *et al.*, 1985). Therefore, the only effect of sonic vibration is to increase the number of filaments without affecting the rates of monomer association to and dissociation from filaments ends.

When sonic vibration is applied to a solution of F-actin at steady state in the presence of ATP, the observed behavior of F-actin is strikingly different: fragmentation is accompanied by a rapid, partial depolymerization to a new steady state (Pantaloni *et al.*, 1984). Further examination of this phenomenon showed that the extent of depolymerization was a function of the regime of fragmentation imposed by the sonicator, i.e., of the number of ends maintained in solution. The fragmentation can actually be controlled at will using a time controller attached to the sonicator, which allows application of sonication periodically for short periods (say, 0.5 sec) separated by variable time intervals. The shorter the interval between two sonications, the smaller the average size of the fragments generated, i.e., the higher the number of filaments. Indeed, polymerization under sustained sonication can be understood as a polymerization with constant filament length, as opposed to seeded polymerization which develops with a constant number of filaments. Very simply, filaments are fragmented when their length exceeds a certain size limit, so that a parameter similar to a "generation time period" can be defined as for bacterial growth. Consequently, it can be demonstrated that the polymerization curve under continuous sonication is symmetric with respect to the point of half polymerization, and can be described by the following equation (Carlier *et al.*, 1985) :

$$\ln [C_0 - C(t)/(C(t) - C_c)] = (k_+/m)(C_0 - C_c)(t - t_{1/2}) \tag{4}$$

In the above equation, C_0, C_c and $C(t)$ are the total actin concentration, the critical concentration and the monomer concentration at time t, respectively. k_+ is the rate constant for monomer association with filament ends, and m is the average number of subunits of sonicated filaments (40-60 subunits). Polymerization under sonication of ADP-actin is adequately described by the above equation. In the presence of ADP, the same monomer-polymer equilibrium is reached with or without sonication, and the same critical concentration can be determined over a range of ADP-actin concentrations.

The situation is quite different when actin is polymerized under sonication in the presence of ATP. In this case, the polymerization curve cannot be described by equation (4). At high actin concentration, overshoot polymerization kinetics were observed, with a maximum and subsequent decrease to a lower stable plateau (Carlier *et al.*, 1985). The final amount of polymer was the same as that obtained when sonication was applied to F-actin that had polymerized spontaneously without sonication. Conversely, when sonication was stopped, repolymerization accompanied the spontaneous length redistribution to a population of less numerous, longer filaments.

In summary, polymerization of ATP-actin under sonication displays two characteristic deviations from the simple law described by equation (4) and valid only for reversible polymerization. These deviations are: i) overshoot polymerization kinetics; ii) the amount of polymer formed decreases - or the steady-state monomer concentration increases - with the number of filaments. These two features are the direct consequence of ATP hydrolysis accompanying the polymerization of ATP-actin as will be explained below.

Because ATP hydrolysis takes place on F-actin with a delay following the incorporation of ATP-subunits, and because in the transient F-ATP state filaments are more stable than in the final F-ADP state, polymerization can be complete, under sonication, within a time short enough for practically all subunits of the filaments to be F-ATP. In a later stage, as P_i is liberated, the F-ADP filament becomes less stable and looses ADP-subunits steadily. The G-ADP-actin liberated in solution is not immediately converted into easily polymerizable G-ATP-actin, because nucleotide exchange is relatively slow on G-actin and is not able by itself to polymerize unless a high concentration (the critical concentration of ADP-actin) is reached. Therefore, G-ADP-actin accumulates in solution. A steady-state concentration of G-ADP-actin is established when the rate of depolymerization of ADP-actin ($k_-[F]$) is equal to the sum of the rates of disappearance of G-ADP-actin via nucleotide exchange and association to filament ends. $[G\text{-}ADP]_{SS}$ within this scheme, is given by the following equation (Pantaloni *et al.*, 1984) :

$$[G\text{-}ADP]_{SS} = \frac{k_{23}[F]}{k_{31} + k_{32}[F]} \tag{5}$$

where k_{23} and k_{32} are the rate constants for ADP-actin dissociation from and association to filament ends, k_{31} is the rate constant for nucleotide exchange on G-actin, and $[F]$ the number concentration of filaments. At a very high filament concentration $[G\text{-}ADP]_{SS}$ reaches a higher limit, $[G\text{-}ADP]_{SS, \infty}$ equal to k_{23}/k_{32} which is the critical concentration for polymerization of ADP-actin. This point can also be experimentally verified (Pantaloni *et al.* , 1984).

The fact that the concentration of G-actin at steady state in the presence of APT varies with the number of filaments may have a biological significance: indeed in cells, large pools of G-ADP-actin may accumulate in regions where a large number of short filaments exist. This behavior is the direct consequence of two combined features of actin polymerization: the hydrolysis of ATP and the relatively slow rate of ATP exchange for ADP on G-actin.

STEREOCHEMISTRY OF NUCLEOTIDE BINDING TO ACTIN AND TUBULIN: ROLE OF DIVALENT METAL ION IN NUCLEOTIDE BINDING AND HYDROLYSIS

Actin binds ATP very tightly in the presence of a divalent metal ion that can be either Ca^{2-} or Mg^{2-}. The exchange inert analogue of Mg-ATP, β,γ-Cr ATP can displace both tightly bound nucleotide and divalent metal ion from G-actin, leading to the conclusion that the tightly bound metal ion interacts with the β- and γ-phosphate of ATP in the nucleotide site (Valentin-Ranc & Carlier, 1989). The conformation, ability to polymerize and rate of ATP hydrolysis differ when CaATP or MgATP is bound to actin (see Carlier (1991) for a review). In particular, the hydrolysis of MgATP is fast, whereas the hydrolysis of CaATP is slow (Carlier *et al.* , 1986) and corresponds to ATP hydrolysis on divalent cation-free actin (Valentin-Ranc & Carlier, 1991). It appears that only Mg^{2+} is able to play an effective role in catalysis. CrATP has also been useful for probing the stereochemistry of ATP binding. The data showed that the metal-ATP chelate was bound in the Λ configuration, which is confirmed by the three-dimensional structure of actin at atomic resolution (Kabsch *et al.* , 1990). CrATP is hydrolysed on F-actin upon polymerization: the hydrolysis product is $Cr\text{-}ADP\text{-}P_i$ that remains bound to F-actin and P_i is not released. The resulting $F\text{-}CrADP\text{-}P_i$-actin filament shows a high stability, as expected.

After metal-ATP hydrolysis on F-actin, only P_i is released in solution, and the β-monodentate metal-ADP remains bound to F-actin. As expected, the $F\ CrADP\text{-}P_i$ filaments are very stable.

MYOSIN SUBFRAGMENT-1 INDUCED POLYMERIZATION OF G-ACTIN

The myosin head has long been shown to induce polymerization of G-actin, even in low ionic strength buffers, into decorated F-actin-S$_1$ filaments that exhibit the classical "arrowhead" structure (Miller *et al.*, 1988 and older references therein). The molecular mechanism of this polymerization process however is unknown.

In an effort to understand how actin-actin interactions might be affected by the binding of the myosin head, and in addition to get some more insight into the nature of the actin-myosin interface, we have investigated the nature of the kinetic actin-myosin intermediates involved in the process of S$_1$-induced polymerization of G-actin. For this purpose, a variety of fluorescent probes (pyrene, NBD, AEDANS) have been covalently attached to the C-terminus of G-actin to probe the G-actin-S$_1$ interaction under conditions of tightest binding, i.e. in the absence of ATP.

a)   Myosin subfragment-1 interacts with 2 G-actin molecules.

The change in intensity of pyrenyl-actin fluorescence (Valentin-Ranc *et al.* , 1991) as well as the change in anisotropy of fluorescence of AEDANS-labeled G-actin (Valentin-Ranc and Carlier, 1992) upon addition of increasing amounts of S$_1$ both yielded titration curves incompatible with the formation of the 1:1 G-actin-S$_1$ complex that was initially proposed (Chaussepied and Kasprzak, 1989). Instead, the binding curves were compatible with the formation of a G$_2$S ternary complex. The S$_1$A$_1$ isomer of S$_1$ showed a higher affinity than S$_1$A$_2$ in this complex. It is plausible that the two G-actin molecules, in the G$_2$S complex, have the same orientation with respect to S$_1$, as the two actin subunits that appear to interact with the myosin head in the rigor state (Milligan *et al.*, 1990), i.e. the two actin monomers in contact via longitudinal bonds along the long pitch helix of the actin filament. Within this view, it is expected that subdomain-2 of one actin molecule in G$_2$S is in contact with subdomain-1 of the other G-actin molecule. Conformational changes in subdomain-2 can actually be monitored by limited proteolysis. Using subtilisin, α-chymotrypsin, trypsin and ArgC protease, we could demonstrate that, consistent with the above hypothesis, binding of S$_1$ to G-actin induces the same changes in subdomain-2 than the G $\rightarrow$ F transition (Fievez and Carlier, 1993). In particular, a new cleavage site (Arg 39-His 40) for ArgC, which is protected in the G-actin conformation, is exposed in G$_2$S and in F-actin or F-actin-S$_1$ conformations.

Covalent crosslinking is traditionally a useful tool to monitor the actin-S$_1$ interface (Audemard *et al.*, 1988 for review). Covalent crosslinking of G-actin-S$_1$ complexes using the zero-length crosslinker EDC revealed that the G-actin-S$_1$ electrostatic close contacts were very similar if not identical to the F-actin-S$_1$ contacts in the rigor filament (Combeau *et al.* 1992). As in F-actin-S$_1$ complexes, only one G-actin could be crosslinked to S$_1$. On the other hand, the main difference between G-actin-S$_1$ and F-actin-S$_1$ is the proximity of the actin C-terminal cys374 to S$_1$ in G-actin-S$_1$ but not in F-actin-S$_1$. Using pPDM (spanning 10 Å), or by photoirradiation of benzophenone-G-actin (prepared by reacting benzophenone maleimide with actin) in complex S$_1$, a 1:1 crosslinked complex of apparent molecular mass 195 KDa in SDS PAGE was obtained. The nature of the aminoacid of S$_1$ that can be crosslinked to G-actin is currently under investigation. This result shows that upon polymerization of actin, a change occurs in the environment of the C-terminal segment of actin; it is known that in the F-actin (and F-actin-S_1) state, cys374 can be crosslinked by pPDM to lys191 of the adjacent actin subunit along the short pitch helix (Elzinga and Phelan, 1984).

b) Oligomers of G-actin and S_1 are the second intermediates in F-acto-S_1 assembly

Light scattering as well as pyrenyl actin or NBD-actin fluorescence changes are convenient to monitor the process of S_1-induced polymerization of G-actin. We have shown that actin-S_1 oligomers form rapidly (within 5s) following formation of GS and G_2S complexes. Analysis of fluorescence data shows that the actin:S_1 molar ratio is 2:1 in these oligomers. The fluorescence of NBD-G-actin, which is not modified upon formation of GS and G_2S complexes is increased ~ 2-fold in the oligomers. The results indicate that oligomers are assembled by condensation of G_2S units, and that new actin-actin interactions, in which hydrophobic contacts are involved, are formed upon oligomer assembly. It is proposed that these actin-actin interactions correspond to the lateral bonds between actin subunits in the filament, along the short pitch helix. Both light scattering and anisotropy of fluorescence measurements indicate that oligomers contain 2-4 G_2S units only. Formation of the decorated filament, in which the actin:S_1 molar ratio is 1:1, requires further endwise condensation of oligomers which results in the creation of new S_1 binding sites, due to the formation of new actin-actin longitudinal bonds.

Increased binding of S_1 presumably accompanies the increase in stability of the final F-actin-S_1 product. The kinetic analysis of these elementary steps leading to the decorated filaments is currently underway.

In conclusion, the kinetics of F-actin-S_1 assembly from G-actin and S_1 does not take place, in a low ionic strength medium, via nucleation of actin filaments followed by S_1 binding, but involves condensation of high affinity $(\text{G-actin})_2\, S_1$ complexes rapidly preformed in solution. Assembly of F-actin-S_1 in the presence of $S_1 \geq$ G-actin, is a quasi-irreversible process. This mechanism is therefore different from the assembly of F-actin filaments, which is characterized by the initial, energetically unfavorable formation of a small number of nuclei representing a minute fraction of the population of actin molecules, followed by endwise elongation from G-actin subunits.

REFERENCES

Bigay, J., Deterre, P., Pfister, C., and Chabre, M., 1987, Fluoride complexes of aluminium or beryllium act on G-proteins as reversibly bound analogues of the γ-phosphate of GTP, *EMBO J.* 6:2907.

Carlier, M.-F., 1987, Measurement of P_i dissociation from actin filaments following ATP hydrolysis using an enzyme-linked assay. *Biochem. Biophys. Res. Comm.* 143:1069.

Carlier, M.-F., Pantaloni, D., and Korn, E.D., 1984, Evidence for an ATP cap at the ends of actin filaments and its regulation of the F-actin steady state, *J. Biol. Chem.* 259:9983.

Carlier, M.-F., Pantaloni, D., and Korn, E.D., 1985, Polymerization of ADP-actin and ATP-actin under sonication and characteristics of the ATP-actin equilibrium polymer, *J. Biol. Chem.* 260:6565.

Carlier, M.-F., and Pantaloni, D., 1986, Kinetic evidence for F-ADP-P_i as a major transient in polymerization of ATP-actin, *Biochemistry* 25:7789.

Carlier, M.-F., Pantaloni, D., and Korn, E.D., 1986, The effects of Mg^{2+} at the high-affinity and low-affinity sites on the polymerization of actin and associated ATP hydrolysis, *J. Biol. Chem.* 261:10785.

Carlier, M.-F., Pantaloni, D., and Korn, E.D., 1987, The mechanisms of ATP hydrolysis accompanying the polymerization of Mg-actin and Ca-actin, *J. Biol. Chem.* 262:3052.

Carlier, M.-F., Pantaloni, D., Evans, J.A., Lambooy, P.K., Korn, E.D., and Webb, M.R., 1987, The hydrolysis of ATP that accompanies actin polymerization is essentially irreversible, *FEBS Lett.* 235:211.

Carlier, M.-F., and Pantaloni, D., 1988, Binding of P_i to F-ADP-actin and characterization of the F-ADP-P_i filament, *J. Biol. Chem..* 263:817.

Carlier, M.-F., 1989, Role of nucleotide hydrolysis in the dynamics of actin filaments and microtubules, *Int. Rev. Cytol.* 115:139.

Carlier, M.-F., 1991, Actin : protein structure and filament dynamics, *J. Biol. Chem.* 266:1.

Chaussepied, P., and Kasprzak, A.A. 1989, Isolation and characterization of the G-actin-myosin head complex, *Nature* 342:950.

Combeau, C., and Carlier, M.-F., 1988, Probing the mechanism of ATP hydrolysis on F-actin using vanadate and the structural analogs of phosphate BeF_3^- and AlF_4^-, *J. Biol. Chem.* 263:17429.

Combeau, C., and Carlier, M.-F., 1989, Characterization of the aluminium and beryllium fluoride species bound to F-actin and microtubules at the site of the γ-phosphate of the nucleotide, *J. Biol. Chem.* 264:19017.

Combeau, C., Didry, D., and Carlier, M.-F., 1992, Interaction between G-actin and myosin subfragment-1 probed by covalent crosslinking, *J. Biol. Chem.* 267:14038.

Dupuis, A., Israel, J.-P., and Vignais, P.V., 1989, Direct identification of the fluoroalumnate and fluoroberyl late species responsible for inhibition of the mitochondrial F_1-ATPase, *FEBS Lett.* 255:47.

Elzinga, M., and Phelan, J.J., 1984, F-actin is intermolecularly crosslinked by NN'phenylenedimaleimide through lysine 191 and cysteine 374. *Proc. Nat. Acad. Sci. USA* 81:6599.

Fievez, S., and Carlier, M.-F., 1993, Conformational changes in subdomain-2 of G-actin upon polymerization into F-actin and upon binding myosin subfragment-1, *FEBS Lett.* 316:186.

Goodno, C.C., 1979, Inhibition of myosin ATPase by vanadate ion. *Proc. Natl. Acad. Sci. USA* 76:2620.

Hibberd, M.G., and Trentham, D.R., 1986, Relationships between chemical and mechanical events during muscular contraction. *A. Rev. Biophys. Chem.* 15:119.

Kabsch, W., Mannherz, H.G., Suck, D., Pai, E.F., and Holmes, K.C., 1990, Atomic structure of the actin: DNase I complex. *Nature Lond.* 347:37.

Korn, E.D., Carlier, M.-F., and Pantaloni, D., 1987, Actin polymerization and ATP hdyrolysis. *Science, Wash.* 238:638.

Miller, L., Phillips, M., and Reisler, E., 1988, Polymerization of G-actin by myosin subfragment-1, *J. Biol. Chem.*. 263:1996.

Milligan, R.A., Whittaker, M., and Safer, D., 1990, Molecular structure of F-actin and location of surface binding sites, *Nature* 248:217.

Mornet, D., Bertrand, R., Pantel, P., Audemard, E., and Kassab, R. 1981, Structure of the actin-myosin interface, *Nature* 292:301.

Oosawa, F., and Asakura, S., 1975, Thermodynamics of the polymerization of protein, Academic Press, London.

Orlova, A., and Egelman, E.H., 1992, Structural basis for the destabilization of F-actin by phosphate release following ATP hydrolysis, *J. Mol. Biol.* 227:1043.

Pantaloni, D., Carlier, M.-F., Coué, M., Lal, A.A., Brenner, S.L., and Korn, E.D., 1984, The critical concentration of actin in the presence of ATP increases with the number of filaments and approaches the critical concentration of ADP-actin, *J. Biol. Chem.*, 259:6274.

Phan, B., and Reisler, E., 1992, Inhibition of myosin ATPase by beryllium fluoride, *Biochemistry* 31:4787.

Valentin-Ranc, C., and Carlier, M.-F., 1989, Evidence for the direct interaction between tightly bound divalent metal ion and ATP on actin. Binding of the Λ isomers of βγ-bidentate CrATP to actin. *J. Biol. Chem.* 264:20871.

Valentin-Ranc, C., and Carlier, M.-F., 1991, Role of ATP-bound divalent metal ion in the conformation and function of actin. *J. Biol. Chem.* 266:7668.

Valentin-Ranc, C., and Carlier, M.-F., 1992, Characterization of oligomers as kinetic intermediates in myosin subfragment-1 induced polymerization of G-actin, *J. Biol. Chem.* 267:21543.

Valentin-Ranc, C., Combeau, C., Carlier, M.-F., and Pantaloni, D., 1991, Myosin subfragment-1 interacts with two G-actin molecules in the absence of ATP. *J. Biol. Chem.* 266:17872.

ACTIN-ASSOCIATED PROTEINS AND CONTROL OF FILAMENT-BASED ASSEMBLY-DISASSEMBLY

STRUCTURAL REQUIREMENTS OF TROPOMYOSIN FOR BINDING

TO FILAMENTOUS ACTIN

Sarah E. Hitchcock-DeGregori

Department of Neuroscience and Cell Biology
Robert Wood Johnson Medical School
Piscataway, NJ 08854

INTRODUCTION

Tropomyosin is an actin binding protein found in virtually all eucaryotic cells. Since its early discovery (Bailey, 1948), the actin binding properties and regulatory functions of tropomyosin have been extensively investigated. In addition, tropomyosin has served as a prototype for the structure of α-helical coiled-coil proteins (Cohen and Parry, 1990). The recent recognition of the diversity of tropomyosins in different cell types and discovery of the relationship between cell shape and isoform expression are indicative of a fundamental role for tropomyosin in the actin cytoskeleton. There is a need for a better understanding of structure-function relationships in this protein.

Tropomyosins form a family of highly-conserved proteins in which diversity is achieved through the existence of different genes and alternative splicing of the transcripts of those genes (reviewed by Lees-Miller and Helfman, 1991). The best-studied gene is that which encodes striated muscle α-tropomyosin, the predominant form in many fast skeletal muscles and in cardiac muscle of small mammals. The same gene gives rise to nine different isoforms as a consequence of alternative splicing of exons encoding regions at or near the N-terminus, in the middle of the molecule, and at the C-terminus. These tropomyosins differ in actin affinity and end-to-end association (eg. Matsumura and Yamashiro-Matsumura, 1985) and have tissue-specific distributions (Lees-Miller and Helfman, 1991).

A common function of tropomyosins is the ability to bind cooperatively to F-actin (Yang et al., 1979). The binding of an isolated tropomyosin to actin is very weak, with the observed overall high affinity a consequence of cooperativity between tropomyosins (Wegner, 1979; Hill et al., 1992). Image analysis has shown that tropomyosin coiled coils are aligned end-to-end (N-terminus to C-terminus) in the grooves of the helical actin filament, as modeled in Figure 1 (O'Brien et al., 1971; Milligan et al., 1990). The molecular length of sarcomeric tropomyosins is such that one tropomyosin molecule spans the length of seven actin monomers in the filament.

Actin: Biophysics, Biochemistry, and Cell Biology
Edited by J.E. Estes and P.J. Higgins, Plenum Press, New York, 1994

The extreme ends form a complex in tropomyosin crystals (Cohen et al., 1971). Interaction between the N- and C-terminal ends has been widely proposed to be responsible for the cooperative actin binding, an idea that is questioned by more recent work to be discussed in this paper.

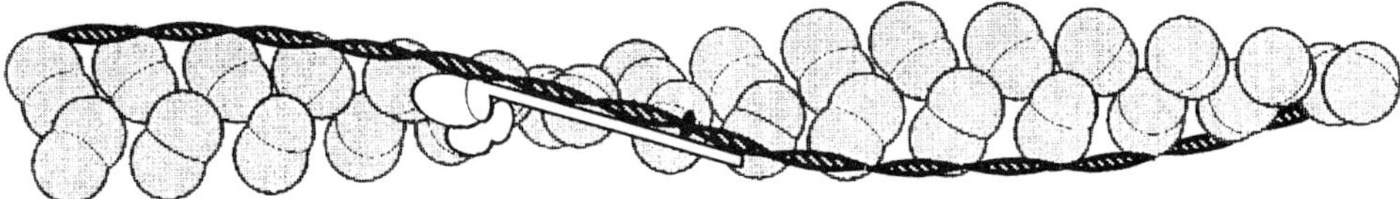

Figure 1. Model of a thin filament showing actin, tropomyosin associated end-to-end spanning the length of seven actin monomers, and troponin. For simplicity, the drawing shows the troponin-tropomyosin only on one side of the actin helix. Adapted from Phillips et al., 1986. Reprinted with permission from S. E. Hitchcock-DeGregori, Cell Motil. and the Cytoskeleton 14:12-20 (1989).

When the amino acid sequence of striated muscle tropomyosin was originally published (Stone and Smillie, 1978), analysis of the sequence revealed the presence of sequence periodicities due to the heptapeptide repeat of the hydrophobic residues important for coiled coil formation, a repeat attributed to gene replication, and a repeat that was attributed to the presence of periodic actin binding sites (Hodges et al., 1972; McLachlan et al., 1975; McLachlan and Stewart, 1976). In their analysis of the sequence, McLachlan and Stewart identified repeats of hydrophobic and polar residues in the outer helical positions of the coiled coil. The repeats were sufficiently regular to correspond to actin binding sites, and they suggested there were two sets of seven sites, which they designated α and ß (Figure 2). Phillips considered in addition the helical positions of residues and proposed there was one set of seven sites, corresponding approximately to McLachlan and Stewart's α-sites (Phillips et al., 1986).

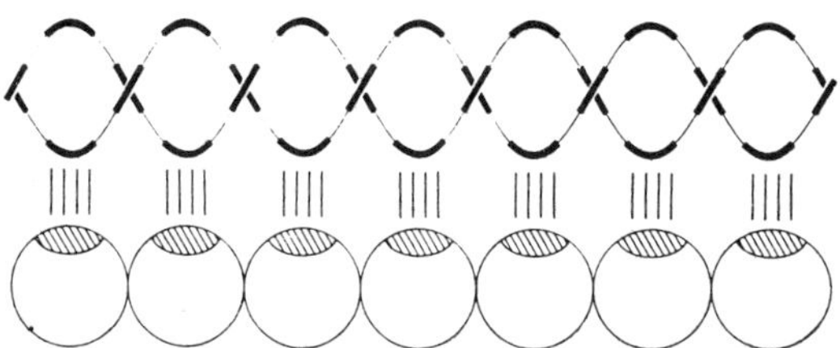

Figure 2. Scheme for the binding of tropomyosin to actin. Each chain in the coiled coil has fourteen repeats that correspond to the two sets of seven α- + ß-sites (McLachlan and Stewart, 1976). Each half-turn of the supercoil has one pair of sites. As a consequence of the symmetry of the coiled coil, each tropomyosin has seven pairs of sites facing the actin filament. Reprinted with permission from M. Stewart and A. D. McLachlan. Nature 257:331-333 (1975).

Tropomyosin has been shown to interact with proteins in addition to actin. The best known are troponin, the protein complex from striated muscles required for Ca^{2+}-dependent regulation (Zot and Potter, 1987); caldesmon, a smooth muscle regulatory protein (Matsumura and Yamashiro, 1993) and tropomodulin, a tropomyosin-binding protein originally isolated from erythrocytes (Fowler, 1987).

The focus of this paper will be to review recent work defining the requirements of tropomyosin for binding to actin. Features of tropomyosin's structure shown to be important for actin binding include the precise spacing of periodic repeats, the sequences of those repeats, and the amino and carboxyl termini. The primary emphasis will be on work using recombinant tropomyosins from our laboratory and others. The binding of troponin to tropomyosin, its effect on actin affinity, and the cooperative interaction between tropomyosin and myosin will not be considered here.

PERIODIC ACTIN BINDING SITES

The structural relationship between tropomyosin and actin in the thin filament implies the presence of periodic actin binding sites. The lengths of tropomyosin correspond to an integral number of actin monomers in the filament: seven in sarcomeric and other 284 residue tropomyosins (McLachlan and Stewart, 1976), six in platelet (Lewis et al., 1983) and most other non-muscle tropomyosins, and five in yeast (references cited in Lees-Miller and Helfman, 1991).

We have tested the hypothesis of periodic actin binding sites by making internal deletions away from the ends, known to be important for binding (as will be discussed later in this paper), and away from the troponin binding site which spans the C-terminal third or more of tropomyosin (Figure 1; White et al., 1987). By making a series of deletions, we have shown that a seven-fold periodicity is important for actin binding, and that a fourteen-fold periodicity is sufficient for binding in the presence of troponin (Hitchcock-DeGregori and Varnell, 1991; Hitchcock-DeGregori and An, unpublished).

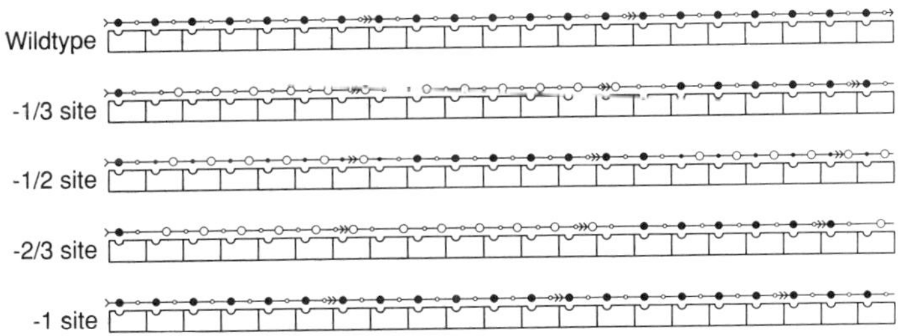

Figure 3. Model of tropomyosin deletion mutants. The design of the experiment was to delete portions of tropomyosin up to a complete putative actin binding site. The drawing shows tropomyosin aligned head-to-tail on the actin filament, wildtype spanning seven actin monomers (illustrated as rectangles). In the wildtype, the α-sites (large dots) are arbitrarily aligned with the binding site on actin. The ß-sites are shown as small dots. Aligned sites are filled, unaligned sites are unfilled. The consequences of the deletions on the alignment of α- and ß-sites are shown in four deletion mutants. Mutants were made in chicken striated muscle α-tropomyosin cDNA (Gooding et al., 1987) using oligonucleotide-directed mutagenesis (Hitchcock-DeGregori and Varnell, 1990). The deletions are in the second actin binding site as follows: 1/3-, res. 47-60; -1/2, res. 47-67; -2/3, res. 47-74, -1, res. 47-88.

The rationale of the approach is diagrammed in Figure 3. The drawing illustrates a filament with actin monomers and associated polar tropomyosins. In wildtype, one tropomyosin spans the length of seven actin monomers in the filament. Neighboring tropomyosins are aligned end-to-end, indicated by arrowheads, along the length of the thin filament. Each tropomyosin molecule has two sets of seven actin binding sites; the large dots correspond to McLachlan and Stewart's α-sites and alternate with small dots corresponding to ß-sites. In the wildtype tropomyosin, the α-sites are shown aligned with the arbitrary binding site on the actin monomer.

We made a series of nested deletions in the chicken striated α-tropomyosin cDNA of the region encoding the second actin binding site (Figure 3). Four deletions were made that correspond to one-third, one-half, two-thirds and one actin binding site. The one site deletion corresponds to McLachlan and Stewart's second α + ß-site and Phillips' second site; the one-half site deletion corresponds to the second α-site. The one-third and two-thirds site deletions are non-integral in that they do not correspond to an α, ß, or α + ß-site. The consequences are that deletion of one site results in a tropomyosin that spans six instead of seven actins in the filament, but the relationship of each site to the filament is unchanged. Deletion of a half-site results in a tropomyosin in which molecules with α-sites aligned alternate with molecules with ß-sites aligned, with 50% of each over the length of the actin filament. In the one-third and two-thirds site deletions, every third molecule is aligned at the α-sites; the intervening two molecules are out of alignment.

The deletions were all multiples of seven amino acids in order to retain the heptapeptide repeat of the coiled coil, the largest deletion being 42 amino acid residues. The mutations were made using oligonucleotide-directed mutagenesis, expressed in E. coli, and purified using conventional methods. Although all these mutants were heat stable, and folded properly, they differed from each other in actin affinity and other tropomyosin functions. We measured the actin binding in the presence of troponin (with Ca^{2+}) since striated α-tropomyosin expressed in E. coli alone binds poorly due to its lack of acetylation (to be discussed later; Heald and Hitchcock-DeGregori, 1988).

Table 1. Actin binding constants and Hill coefficients of wildtype and mutant tropomyosins.[1]

	K_{app} (M^{-1})	Hill Coefficient
Wildtype	5.7 X 10^6	2.7
-1/3 site	< <10^5	n.d.
-1/2 site	5.8 X 10^6	2.7
-2/3 site	< <10^5	n.d.
-1 site	3.3 X 10^6	2.3

[1]Binding was carried out in the presence of troponin and Ca^{2+}. All data except -1/3 site are from Hitchcock-DeGregori and Varnell (1990). The mutants are described in Figure 3.

Table 1 summarizes the actin binding data of these four mutants. Compared to wildtype, it is clear that deletion of one-half or the full second actin binding site results in a small reduction in actin affinity in the presence of troponin without affecting the cooperativity, as reflected in the similar Hill coefficients. A mutant with the third site deleted also binds to actin though the reduction in affinity compared to wildtype is about seven-fold, greater than with deletion of the second site, suggesting that the third site may contribute more to the overall actin affinity than the second site (Hitchcock-DeGregori and An, unpublished results). In contrast, a non-integral deletion (one-third or two-thirds of a site) results in loss of actin affinity within the detection limits of our assays (Table 1). These results show that periodic actin binding sites are important for high affinity binding of tropomyosin to actin. The presence of an integral number of actin binding sites is more important than the total number of periods. Our results imply that each actin binding site on tropomyosin contributes in only a small way to the overall actin affinity. The apparent actin affinity is a consequence of the binding of an isolated tropomyosin to actin and a cooperativity parameter. If each of the seven sites contributed equivalently to the non-cooperative binding constant (Wegner, 1981; Hill et al., 1992), each site would have, on average, an affinity of about 3 M^{-1}, consistent with the reduction in affinity observed when we delete individual actin binding sites.

I return now to the identification of the actin binding sites in the tropomyosin sequence. McLachlan and Stewart (1976) used Fourier transformation to define the α- and ß-sites primarily on the basis of the periods of alternating positively and negatively charged residues in the outer helical positions (b, c, f) of the heptapeptide repeat in the tropomyosin supercoil. (McLachlan et al., 1975, described the positions of amino acid residues in the heptapeptide repeat a-g, with a and d occupying the hydrophobic interface between the α-helices.) While bands of charges are evident, the repeats are poor when examined at the level of actual amino acid sequence. It is not possible to line up the repeats as with the gelsolin superfamily of actin binding proteins such as villin (Bazari et al., 1988) or the sarcomeric protein titin (Labeit et al., 1990). Phillips (Phillips et al., 1986) took the McLachlan and Stewart analysis further and considered the helical (azimuthal) position of the amino acid on the α-helix of the coiled coil, as well as the linear (supercoil) position. He found a seven-fold periodicity

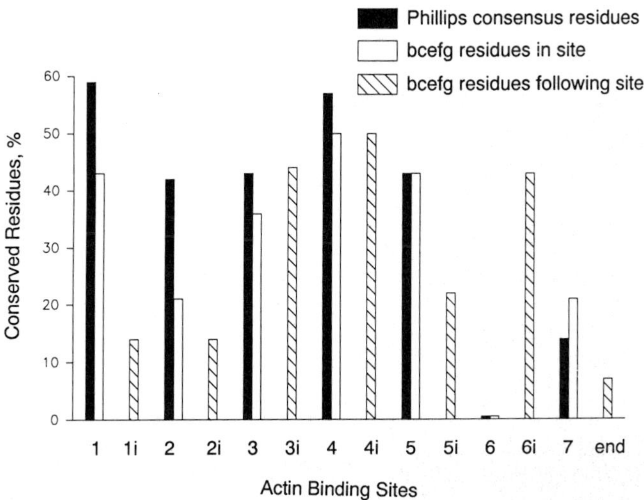

Figure 4. Conservation of residues in Phillips' proposed actin binding sites compared to residues not in the site. The sites 1-7 refer to the seven actin binding sites that correspond approximately to the McLachlan and Stewart (1976) α-sites. 1i-6i and end refer the regions between these sites, corresponding approximately to the ß-sites. The Phillips consensus residues refer to the seven residues in his proposed sites. *bcefg* refer to all the residues in the non-interface positions in the heptapeptide repeat of the coiled coil (McLachlan et al., 1975). The analysis was only in terms of identities. Sixteen sequences were included: chicken, rabbit, rat, human striated α; chicken, rat smooth α; rabbit, rat striated ß; chicken, rat smooth ß, equine platelet, rat TM4, rat TM5b, human TM3; Drosphila TM I, nematode. References for the sequences can be found in Lees-Miller and Helfman (1991). The Phillips consensus residues refer to the 7 residues in his proposed sites.

that corresponds approximately to McLachlan and Stewart's seven α-sites. Further analysis revealed a subset of seven residues that are well-conserved or homologous in the repeats. Phillips has suggested these seven residues define periodic binding sites. (Phillips, personal communication). These charged and non-polar residues form a stripe approximately along the length of each α-site.

Since the sequence of actin is so highly conserved throughout phylogeny, one might postulate that the residues involved in actin binding would also be conserved. With this in mind, I have evaluated the conservation of residues in Phillips' proposed sites among vertebrate and invertebrate tropomyosins, emphasizing 284 residue isoforms. The residues in Phillips' sites were compared to all the non-interface (*bcefg*) residues in and between sites. The analysis included 13 tropomyosins from 7 animal species and 4 vertebrate genes. It did not include exons encoding the N-terminal residues of 246 residue tropomyosins (such as platelet) as well as the non-conventional C-terminal exons found in <u>Drosphila</u> and rat brain.

Figure 4 shows that the site residues are typically about 40-50% conserved (identical) between tropomyosins except for two sites which are poorly conserved, sites 6 and 7. Surprisingly, only two residues each of sites 6 and 7 are in alternatively spliced exons. Site 2, which is about 40% conserved, <u>is</u> encoded by an alternatively spliced exon. Generally, the residues in the site are about as well conserved as the whole region suggesting that there is no strong selective pressure on the proposed site residues. Similarly, the residues in the site may be more or less conserved than the bcefg residues between the sites.

Then, one might ask, do the periods contribute in any specific way to actin affinity, or are they merely spacers to place the ends, or a small number of sites, in the proper relationship to the actin filament? It is possible that the reduction in affinity we observed upon deletion of the second or third site is due to imperfect deletion of a period resulting in inaccurate alignment of the ends.

Results from our laboratory and others, however, suggest that the internal periods contribute in a positive fashion to actin affinity, that they are not merely spacers. Evidence comes from two types of experiments. First, it is well known that

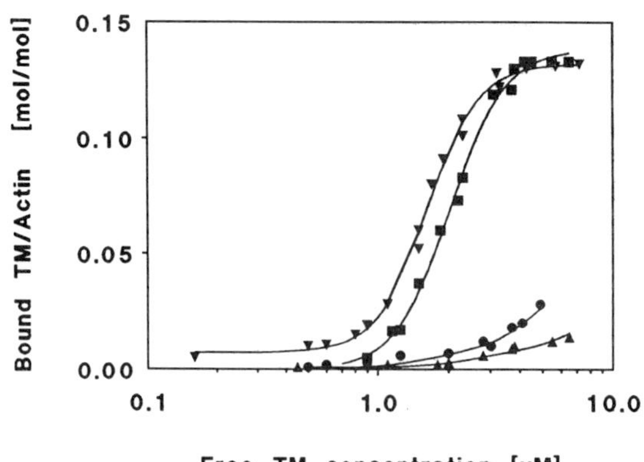

Figure 5. Actin binding of recombinant (unacetylated) rat striated, smooth, and chimeric tropomyosins to actin. These tropomyosins differ due to alternative splicing of the 2nd (res. 39-81) and terminal coding 9th (res. 258-284) exons. The binding of tropomyosin to F-actin was measured by cosedimentation and densitometric analysis of the supernatants and pellets. Squares, smooth TM; circles, striated TM; triangles, TM with smooth exon 2/striated exon 9; inverted triangles, TM with striated exon 2/smooth exon 9. Reprinted with permission from Cho and Hitchcock-DeGregori (1991).

90

tropomyosin isoforms differ in actin affinity. In two instances, it has been possible to relate the differences to specific internal sequences. In the first case, we have analyzed the effect of the 2nd exon encoding res. 39-80, a region that is alternatively spliced in smooth and striated tropomyosins (Ruiz-Opazo and Nadal-Ginard, 1987) on actin affinity (Figure 5; Cho and Hitchcock-DeGregori, 1991). Tropomyosins with the striated 2nd exon have a slightly higher actin affinity. Similarly, when the 6th exon, encoding res. 189-214, is replaced with a form found only in non-muscle cells, there is a two-fold higher actin affinity (Pittinger and Helfman, 1992; Hammell and Hitchcock-DeGregori, unpublished results).

In order to define the specificity for actin binding, we have begun substituting regions of tropomyosin with non-tropomyosin sequences (Hitchcock-DeGregori and An, unpublished results). In our first mutant of this sort we replaced 14 residues of tropomyosin (1/3 of the second actin binding site) with 14 residues of the leucine zipper, GCN4 (Landschulz et al., 1988). Though this is a coiled-coil sequence, it is unrelated to tropomyosin. This change resulted in less than a two-fold reduction in actin affinity, comparable to deletion of the entire second actin binding site. A similar mutation of the third actin binding site resulted in a 2-3-fold reduction in actin affinity. Replacement with 14 residues of site 2 with 14 residues of random coil resulted in loss of actin affinity even though the mutant can form a coiled coil. From these analyses we conclude that the periodic repeats <u>do</u> contribute to the overall actin affinity and function as more than mere spacers. The small affinity attributable to each site may not require strict conservation of sequence. The continuity of the coiled coil along the length appears to be important for actin binding.

IMPORTANCE OF THE ENDS

While the internal actin binding sites contribute in only a small way to the overall actin affinity, it is clear that both the amino- and carboxyl-terminal ends of tropomyosin are of paramount importance. When the sequence of tropomyosin was originally determined, McLachlan and Stewart (1975) proposed that the first and last nine residues formed an overlap region of the C-terminus of one tropomyosin with the N-terminus of the next molecule, a model to accommodate the known interaction between the ends tropomyosin in crystals (Cohen et al., 1971). This overlap region was postulated to be required for the polymerization of tropomyosin <u>in vitro</u> and the cooperative actin binding that results in high actin affinity and head-to-tail alignment of tropomyosin on the thin filament. Results from several laboratories have shown that both ends are crucial for tropomyosin function. We know that the role of the ends must be modified from the original view, even though much remains to be learned.

The amino terminus

The amino terminus of muscle tropomyosins, and other 284 residue isoforms, is the most highly conserved region of the molecule; the first nine residues are identical in <u>Drosophila</u> and man. Our work with recombinant tropomyosins has shown that this region is critical for all functions of striated muscle tropomyosin: actin and troponin binding, head-to-tail association, and regulation (Hitchcock-DeGregori and Heald, 1987; Heald and Hitchcock-DeGregori, 1988; Cho et al., 1990).

Studies with recombinant tropomyosins have shown that N-terminal modification of the initial methionine is important for cooperative binding of striated α-tropomyosin to actin. The first clue to the importance of this common post-translational modification came when we expressed tropomyosin in <u>E. coli</u>. The non-fusion protein, which is unacetylated, bound poorly to actin in the absence of troponin. Interestingly, the lack of acetylation reduces the affinity without significantly altering the cooperativity (Heald and Hitchcock-DeGregori, 1988; Willadsen et al., 1992). However, with 80 residues of a viral protein on the N-terminus, it bound with an affinity and cooperativity similar to that of tropomyosin isolated from muscle (Heald and Hitchcock-DeGregori, 1988). It is now clear that almost any fusion peptide on the N-terminus will allow actin binding, since tropomyosins with 30 residues (Stone and Mendelson, 1989) and as few as 3 residues on the amino terminus will bind (Urbancikova and Hitchcock-DeGregori, unpublished results).

The failure of unacetylated tropomyosin to bind to actin does not appears to be a consequence of the protonated amino terminus. First, it does not bind at pH 9.4, at which it should be more than half deprotonated (Cho et al., 1990). Second, a fusion tropomyosin with Arg next to the Met 1 of tropomyosin binds with high affinity to actin (Urbancikova and Hitchcock-DeGregori, unpublished results). N-terminal acetylation does stabilize synthetic amino-terminal peptides, but it has little effect on the stability of the full-length tropomyosin (Greenfield and Hitchcock-DeGregori, submitted). A fuller understanding of the importance of this common post-translational modification on actin binding will await solution of the structure.

The conserved N-terminal sequence is also important for function. The most dramatic demonstration of its functional significance is that deletion of the first nine residues results in complete loss of striated α-tropomyosin function, even though it folds properly to form a parallel dimer with fairly normal stability (Heald and Hitchcock-DeGregori, 1988; Cho et al., 1990). This mutant had no detectable affinity for actin, even in the presence of troponin (Figure 6), and exhibited no regulatory function. Indications of the importance of this region also came from earlier work which showed that chemical modification of Lys 7 resulted in loss of polymerizability (Johnson and Smillie, 1977).

Our results with N-terminal tropomyosin variants have led us to postulate that the conserved N-terminal region may be an actin binding site that contributes more to the overall affinity than the internal periodic sites discussed earlier (Cho et al., 1990), a conclusion also reached recently by Willadsen et al. (1992) based on a more extensive analysis of the cooperativity of actin binding. The sequence of the conserved amino-terminal region, MDAIKKKMQ, appears to be unrelated to the consensus actin binding site in the gelsolin family of proteins (Bazari et al., 1988), but it is similar to

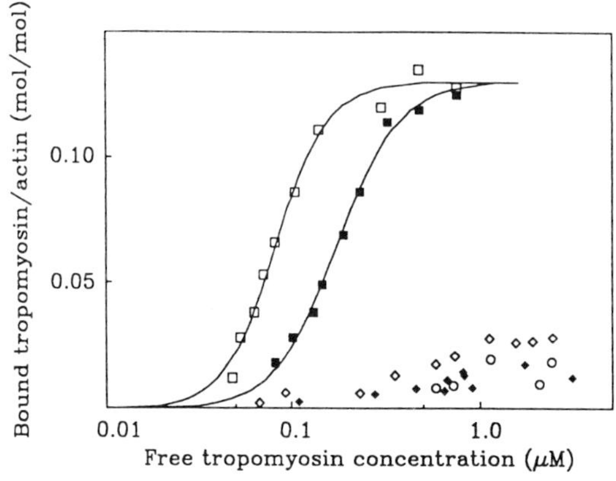

Figure 6. Binding of wildtype and N-terminal deletion mutant (d1-9) to F-actin in the presence of troponin. Squares, wildtype; diamonds, mutant; circles, mutant, no troponin. Filled symbols, +Ca^{2+}; open symbols, +EGTA. Reprinted with permission from Cho et al. (1990).

a sequence (DAIKKK) found in cofilin, a protein that binds stoichiometrically to F-actin (Matsuzaki et al., 1988). Crosslinking experiments, however, have identified a different region of cofilin that binds to actin (Yonezawa et al., 1991), leaving unclear the role of the tropomyosin-like sequence in cofilin.

The carboxyl terminus

The importance of the carboxyl terminus for the function of striated muscle tropomyosin is well established. When the C-terminal 9-11 residues are removed using carboxypeptidase A, tropomyosin does not polymerize, troponin does not induce polymerization, and its affinity for actin is greatly reduced. Troponin, however, increases the affinity of carboxypeptidase-treated tropomyosin for actin, and regulation of the MgATPase with troponin is apparently normal, though myosin Sl-induced cooperativity is reduced (Ueno et al., 1976; Mak and Smillie, 1981; Dabrowska et al., 1983; Walsh et al., 1984; Heeley et al., 1987, 1989). Interestingly, the properties of carboxypeptidase-treated tropomyosin are similar to those of unacetylated tropomyosin produced in $\underline{E.\ coli}$ I have just described: the C-terminal region appears to contribute primarily to actin affinity rather than the cooperativity of binding. It is also worth emphasizing that removal of the amino-terminal nine residues is much more deleterious than removal of the carboxyl-terminal region. If the N- and C-terminal nine residues served only as an overlap region between neighboring molecules, we would expect the deletions to have comparable effects.

Another demonstration of the importance of the C-terminal region of tropomyosin for actin affinity comes from analysis of tropomyosins differing in the ninth exon encoding the C-terminus (res. 259-284). Smooth and striated α-tropomyosins differ in the second and ninth exons (Ruiz-Opazo and Nadal-Ginard, 1987). We have evaluated the contributions of these exons to tropomyosin function by expressing smooth, striated, and smooth-striated chimeric tropomyosins in $\underline{E.\ coli}$ (Cho et al., 1991). Surprisingly, the recombinant smooth α-tropomyosin bound well to tropomyosin, much better than the striated form (Figure 5). Binding was cooperative, even though polymerization was poor. When we analyzed the chimeras, the difference in affinity was attributed nearly entirely to the ninth exon. The ninth exon encodes the C-terminal 25 amino acids. Interestingly, the smooth form of the ninth exon is present in most non-muscle α-tropomyosin isoforms, whereas the striated form is only found in striated muscles and appears to be specialized for Ca^{2+}-independent interaction with troponin on the thin filament (Cho and Hitchcock-DeGregori, 1991).

The whole is less than the sum of the parts.

It is puzzling that smooth muscle α-tropomyosin, and other α-tropomyosin isoforms having the same C-terminal exon, expressed in $\underline{E.\ coli}$ have similar actin affinities to that of striated muscle α-tropomyosin (purified from muscle), even though they are unacetylated (Cho et al., 1991; Pittinger and Helfman, 1992; Hammell and Hitchcock-DeGregori, unpublished results). In other words, the alterations at the two ends do not seem to be additive. This question has been directly addressed by Butters et al. (1993) who showed that the actin affinity of carboxypeptidase-treated striated α-tropomyosin was similar in the acetylated (purified from muscle) or unacetylated (expressed in $\underline{E.\ coli}$) forms. Possibly only one "strong binding" end is necessary: the highly conserved amino terminus may be more important for the actin affinity of striated tropomyosin which has a "weak binding" carboxyl-terminus than it is for smooth tropomyosin and other isoforms with the "strong binding" ninth exon.

RELATIONSHIP BETWEEN THE PERIODIC ACTIN BINDING SITES AND THE ENDS

Together, the results discussed here suggest that each periodic actin binding site contributes in a small way to the overall actin affinity, but that their most important role may be to ensure the proper structural relationship of the N- and C-terminal ends of tropomyosin to each other and to the thin filament. The alterations at the ends of tropomyosin all appear to affect the affinity of tropomyosin for actin without primarily altering the cooperativity of binding (Dabrowoska et al., 1983; Heald and Hitchcock-DeGregori, 1988; Cho et al., 1990; Cho and Hitchcock-DeGregori, 1991; Willadsen et al., 1992).

The source of the cooperativity of tropomyosin binding to actin is not understood. In native thin filaments, the actin is fully saturated with tropomyosin which is aligned head-to-tail along the length of the filament. In addition, from Ohtsuki's work with troponin T antibodies (1979) one may infer that tropomyosins on opposite sides of the filament are aligned in parallel. One of our goals is to understand the molecular basis for this cooperativity.

Since tropomyosin polymerizes at low ionic strength <u>in vitro</u> (Bailey, 1948) and is aligned end-to-end on the thin filament (O'Brien et al., 1971; Milligan et al., 1990), the cooperativity of binding intuitively has been attributed to interaction between the first and last nine-residues of tropomyosin, referred to as the overlap region. Although it is difficult to relate the meaning of the viscosity of tropomyosin alone to the arrangement of tropomyosin on the actin filament (since the structure of the ends is unknown), it is clear that there is no consistent correlation between cooperative actin binding and polymerizability. Interestingly, Graceffa (1992) has shown that the viscosity of smooth muscle tropomyosin is irreversibly reduced upon heating, a common step in the preparation of non-muscle and recombinant tropomyosins. Certain naturally occurring non-muscle tropomyosins, as well as recombinant and chemically modified forms of tropomyosin, bind cooperatively to actin even though they do not polymerize <u>in vitro</u> (e.g., Johnson and Smillie, 1977; Dabrowska et al., 1983; Matsumura and Yamashiro-Matsumura, 1985; Broschat and Burgess, 1986; Hitchcock-DeGregori and Heald, 1987; Cho and Hitchcock-DeGregori, 1991). This raises the possibility that the cooperativity between tropomyosin molecules is somehow generated via the actin filament itself (Heeley et al., 1987; Heald and Hitchcock-DeGregori, 1988; Cho et al., 1990; Willadsen et al., 1992; Butters et al., 1993).

ACKNOWLEDGMENTS

I wish to thank my co-workers, collaborators and colleagues who have helped to make this work interesting and fun. I also thank George Phillips for discussions and for sending me his unpublished analysis of the actin binding sites. The research has been supported by the American Heart Association, American Heart Association N. J. Affiliate, Muscular Dystrophy Association and National Institutes of Health.

REFERENCES

Bailey, K., 1948, Tropomyosin: A new asymmetric protein component of the muscle fibril, *Biochem. J.* 43:271-279.

Bazari, W. L., Matsudaira, P., Wallek, M., Smeal, T., Jakes, R., and Ahmed, Y., 1988, Villin sequence and peptide map identify six homologous domains. *Proc. Natl. Acad. Sci. USA.* 85:4986-4990.

Broschat, K.O., and Burgess, D. R., 1986, Low M_r tropomyosin isoforms from chicken brain and intestinal epithelium have distinct actin binding properties, *J. Biol. Chem.* 28:13350-13359.

Butters, C. A., Willadsen, K. W., and Tobacman, L. S., 1993, Cooperative interactions between adjacent troponin-tropomyosin complexes may be transmitted through the actin filament, *J. Biol. Chem.* 268:15565-15570.

Cohen, C., Caspar, D. L. D., Parry, D. A. D., and Lucas, R., 1971, Tropomyosin crystal dynamics. *Cold Spring Harbor Symp. Quant. Biol.* 36:205-216.

Cohen C., and Parry, D. A. D., 1990, α-Helical coiled coils and bundles :how to design an α-helical protein, *Proteins: Str., Funct. Gen.* 7:1-15.

Cho, Y.-J., and Hitchcock-DeGregori, S.E., 1991, Relationship between alternatively spliced exons and functional domains in tropomyosin, *Proc. Natl. Acad. Sci. U.S.A.* 88:10153-10157.

Cho, Y.-J., Liu, J., and Hitchcock-DeGregori, S. E., 1990, The amino terminus of muscle tropomyosin is a major determinant for function. *J. Biol. Chem.* 265:538-545.

Dabrowska, R., Nowak, E., and Drabikowski, W., 1983, Some functional properties of nonpolymerizable and polymerizable tropomyosin, *J. Muscle Res. and Cell Motil.* 4:143-161.

Fowler, V. M., 1987, Identification and purification of a novel M_r 43,000 tropomyosin-binding protein from human erythrocyte membranes, *J. Biol. Chem.* 262:12792-12800.

Gooding, C., Reinach, F. C., and MacLeod, A. R., 1987, Complete nucleotide sequence of the fast-twitch isoform of chicken skeletal muscle α-tropomyosin, *Nuc. Acids Res.* 15:8105.

Graceffa, P., 1992, Heat-treated smooth muscle tropomyosin, *Biochim. Biophys. Acta.* 1120:205-207.

Heald, R. W., and Hitchcock-DeGregori, S. E., The structure of the amino terminus of tropomyosin is critical for binding to actin in the absence and presence of troponin, *J. Biol. Chem.* 263:5254-5259.

Heeley, D. H., Golosinska, K., and Smillie, L. B., 1987, The effects of troponin T fragments T1 and T2 on the binding of nonpolymerizable tropomyosin to F-actin in the presence and absence of troponin I and troponin T, *J. Biol. Chem.* 262:9971-9978.

Heeley, D.H., Smillie, L. B., and Lohmeier-Vogel, E.M., 1989, Effects of deletion of tropomyosin overlap on regulated actomyosin subfragment 1 ATPase, *Biochem. J.* 258:831-836.

Hill, L. E., Mehegan, J. P., Butters, C. A., and Tobacman, L. S., 1992, Analysis of troponin-tropomyosin binding to actin. *J. Biol. Chem.* 267:16106-16113.

Hitchcock-DeGregori, S.E., and Varnell, T. A., 1990, Tropomyosin has discrete actin-binding sites with sevenfold and fourteenfold periodicities, *J. Mol. Biol.* 214:885-896.

Hodges, R. S., Sodek, J., Smillie, L. B., and Jurasek, L., 1972, Tropomyosin: Amino acid sequence and coiled-coil structure, *Cold Spring Harbor Symp. Quant. Biol.* 37:299-310.

Johnson, P., and Smillie, L. B., 1977, Polymerizability of rabbit skeletal tropomyosin: effects of enzymatic and chemical modifications, *Biochemistry.* 16:2264-2269.

Labeit, S., Barlow, D. P., Gautel, M., Gibson, T., Holt, J., Hsieh, C.-L., Francke, U., Leonard, K., Wardale, J., Whiting, A., and Trinick, J., 1990, A regular pattern of two types of 100-residue motif in the sequence of titin, *Nature* 345:273-276.

Landschulz, W. H., Johnson, P. F., and McKnight, S. L., 1988, The leucine zipper: a hypothetical structure common to a new class of DNA binding proteins, *Science* 240:1759-1764.

Lees-Miller, J. P., and Helfman, D.M., 1991, The molecular basis for tropomyosin isoform diversity, *Bioessays* 13:429-437.

Lewis, W. G., Cote, G. P., Mak A. S., and Smillie, L. B., 1983, Amino acid sequence of equine platelet tropomyosin, *FEBS Letts* 156:269-273.

Mak, A. S., and Smillie, L. B., Non-polymerizable tropomyosin: preparation, some properties and F-actin binding, *Biochem. Biophys. Res. Commun.* 101:208-214.

Matsumura F., and Yamashiro-Matsumura, S., 1985, Purification and characterization of multiple isoforms of tropomyosin from rat cultured cells, *J. Biol. Chem.* 260:13851-13859.

Matsumura F., and Yamashiro, S., 1993, *Caldesmon. Curr. Opin. Cell Biol.* 5:70-76.

Matsuzaki, F., Matsumoto, S., Yahara, I., Yonezawa, N., Nishida, E., and Sakai, H., 1988, Cloning and characterization of porcine brain cofilin cDNA, *J. Biol. Chem.* 263:11564-11568.

McLachlan, A. D., and Stewart, M., 1975, Tropomyosin coiled-coil interactions: Evidence for an unstaggered structure, *J. Mol. Biol.* 98:293-304.

McLachlan, A. D., Stewart, M., and Smillie, L. B., 1975, Sequence repeats in α-tropomyosin. *J. Mol. Biol.* 98:281-291.

McLachlan, A. D., and Stewart, M., 1976, The 14-fold periodicity in α-tropomyosin and the interaction with actin, *J. Mol. Biol.* 103:271-298.

Milligan, R. A., Whittaker, M., and Safer, D., 1990, Molecular structure of F-actin and location of surface binding sites, *Nature.* 348:217-221.

O'Brien, E. J., Bennett, P. M., and Hanson, J., 1971, Optical diffraction studies of myofibrillar structure, *Phil. Trans. Roy. Soc. London B.* 261:201-208.

Ohtsuki, I., 1979, Molecular arrangement of troponin T in the thin filament, *J. Biochem.(Tokyo).* 86:491-497.

Phillips, Jr., G. N., Fillers J. P., and Cohen, C., Tropomyosin crystal structure and muscle regulation, *J. Mol. Biol.* 192:111-131.

Pittinger, M. F., and Helfman, D. M. 1992, In vitro and in vivo characterization of four fibroblast tropomyosins produced in bacteria: TM-2, TM-3, TM-5a, and TM-5b are co-localized in interphase fibroblasts, *J. Cell Biol.* 118:841-858.

Ruiz-Opazo, N., and Nadal-Ginard, B., α-Tropomyosin gene organization, *J. Biol. Chem.* 262:4755-4765.

Stone, D., and Smillie, L. B., 1978, Amino acid sequence of rabbit skeletal α-tropomyosin. The NH_2-terminal half and complete sequence, *J. Biol. Chem.* 253:1137-1148.

Stone, D. B., and Mendelson, R. A., A tropomyosin fusion protein and its factor X_a cleavage product regulate acto-S1 MgATPase in the presence of troponin, *Biophys. J.* 55:277a.

Ueno, H., Tawada, Y., and Ooi, T., Properties of non-polymerizable tropomyosin obtained by carboxypeptidase A digestion, *J. Biochem.* 80:283-290.

Walsh, T. P., Trueblood, C. E., Evans, R., and Weber, A., 1985, Removal of tropomyosin overlap and the co-operative response to increasing calcium concentrations of the acto-subfrabment-1 ATPase, *J. Mol. Biol.* 182:265-269.

Wegner, A., 1981, Equilibrium of the actin-tropomyosin interaction, *J. Mol. Biol.* 131:839-853.

White, S. P., Cohen, C., and Phillips, Jr., G.N., 1987, Structure of co-crystals of tropomyosin and troponin. Nature 325:826-828.

Willadsen, K. A., Butters, C. A., Hill, L. E., and Tobacman, L. S., 1992, Effects of the amino-terminal regions of tropomyosin and troponin T on thin filament assembly, *J. Biol. Chem.* 267:23747-23752.

Yang, Y., Korn, E. D., and Eisenberg, E., 1979, Cooperative binding of tropomyosin to muscle and Acanthamoeba actin, *J. Biol. Chem.* 254:7137-7140.

Yonezawa, N., Nishida, E., Iida, K., Kumagai, H., Yahara, I., and Sakai, H., 1991, Inhibition of actin polymerization by a synthetic dodecapeptide patterned on the sequence around the actin-binding site of cofilin, *J. Biol. Chem.* 266:10485-10489.

Zot, A. S., and Potter, J. D., 1987, Structural aspects of troponin-tropomyosin regulation of skeletal muscle contraction, *Annu. Rev. Biophys. Chem.* 16:535-559.

ACTIN-GELSOLIN INTERACTION

Albrecht Wegner[1], Klaus Aktories[2], Andrea Ditsch[1], Ingo Just[2],
Beate Schoepper[1], Norma Selve[1] and Michaela Wille[2]

[1]Institute of Physiological Chemistry,
Ruhr-University, D 4630 Bochum, F. R. G.
[2]Institute of Pharmacology and Toxicology
University of the Saarland, D 6650 Homburg/Saar, F. R. G.

INTRODUCTION

Gelsolin is an actin-polymerization-regulating protein that has been isolated from many types of cells.[1,2] Gelsolin affects actin polymerization in different ways. Gelsolin can fragment actin filaments. Gelsolin binds to the barbed ends of the fragmented actin filaments. The pointed ends of actin filaments remain free for depolymerization and polymerization. Gelsolin nucleates actin filaments to polymerize towards the pointed ends. Furthermore, gelsolin forms complexes with one or two actin molecules. These gelsolin-actin complexes lose the ability to fragment actin filaments. They cap actin filaments by binding to the barbed ends to inhibit polymerization and depolymerization of these ends (Fig. 1).[1-4] These interactions of gelsolin with actin filaments are mainly regulated by the second messengers calcium and phosphatidyl inositol phosphates.[1,5] Because of the versatility of regulation and interactions with actin and because of its ubiquity in eukaryotic cells gelsolin has attracted our interest. We investigated the equilibrium and the rates of various interactions of gelsolin with actin filaments.

FORMATION OF 1:1 AND 1:2 GELSOLIN-ACTIN COMPLEXES

The time course of the formation of the gelsolin-actin complex in the presence of micromolar concentrations of Ca^{2+} was followed by the increase of the fluorescence intensity of a fluoresence label bound to actin (Fig. 2). The time course of this assembly reaction could be quantitatively interpreted by a model in which one actin molecule binds slowly to gelsolin in a rate-determining step and subsequently a second actin molecule is bound considerably more rapidly. The rate of binding of the first actin

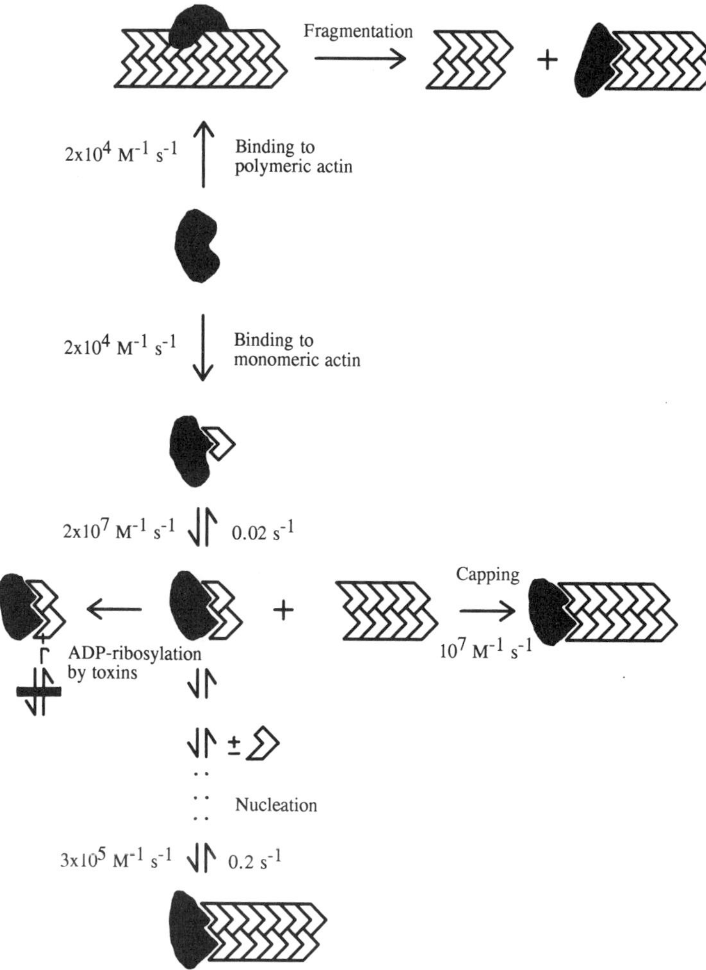

Fig. 1. Reactions of gelsolin with actin.

molecule (A) to gelsolin (G) was found to be remarkably low and to depend on the pH. The rate constants $k_1{}^+$ of formation of the 1:1 gelsolin-actin complex (GA) ranged from 1.5×10^4 M^{-1} s^{-1} at pH 8 to 7×10^4 M^{-1} s^{-1} at pH 6.[6]

The rate and equilibrium constant of association of fluorescently labeled actin monomers with 1:1 gelsolin-actin complex were measured by fluorescence stopped flow experiments (Fig. 3). The rate constant $k_2{}^+$ of formation of the 1:2 gelsolin-actin complex from 1:1 gelsolin-actin complex and actin was found to be about 2×10^7 M^{-1} s^{-1} under conditions where gelsolin binds Ca^{2+}. Thus, the second actin molecule binds 1000-fold faster to gelsolin than the first actin molecule.[7]

The rate of dissociation of one actin molecule from the 1:2 gelsolin-actin complex was determined by exchange of actin for fluorescently labeled actin. For these experiments gelsolin was isolated from chicken gizzard or plasma. The rate constant $k_2{}^-$ of dissociation of one actin molecule from the 1:2 gelsolin-actin complex was found to be about 0.02 s^{-1} (Fig. 1). Thus, the equilibrium constant K_2 for association of actin with 1:1 gelsolin-actin complex can be calculated to be in the range of 10^9 M^{-1} ($K_2 = k_2{}^+/k_2{}^-$). The rate of dissociation of actin from 1:2 gelsolin-actin complex in the absence of Ca^{2+} ions was measured following addition of EGTA to 1:2 gelsolin-actin

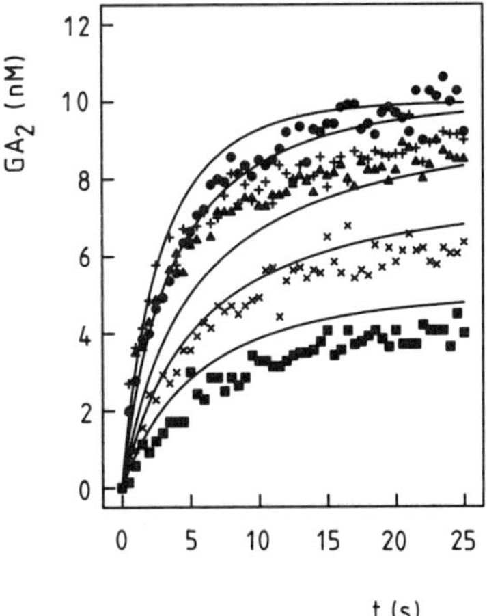

Fig. 2. Rate of formation of the 1:2 gelsolin-actin complex from gelsolin and actin at pH 7. GA_2, concentration of the 1:2 gelsolin-actin complex. 50 nM gelsolin were mixed with the following concentrations of actin: ■, 25 nM; ▲, 62.5 nM; x, 100 nM, o, 137.5 nM. The continuous lines were calculated for the rate constant of binding of actin to gelsolin $k_1^+ = 2.5 \times 10^4$ M^{-1} s^{-1}.

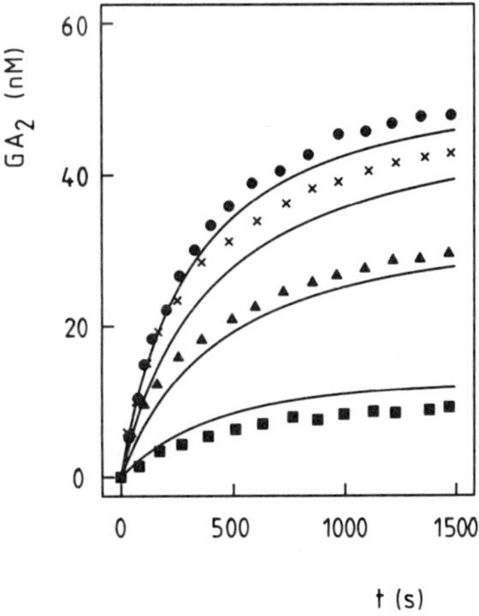

Fig. 3. Rate of formation of the 1:2 gelsolin-actin complex from 1:1 gelsolin-actin complex and actin (stopped flow experiments). GA_2, concentration of the 1:2 gelsolin-actin complex. 10 nM gelsolin-actin complex were mixed with the following concentrations of actin: ■, 5 nM; x, 7.5 nM ; ▲, 10 nM; +, 15 nM, o, 20 nM. The continuous lines were calculated for the rate constant of binding of actin to 1:1 gelsolin-actin complex $k_2^+ = 2 \times 10^7$ M^{-1} s^{-1}.

complex. The rate of dissociation was found to be about 0.02 s^{-1} similarly to the rate in the presence of Ca^{2+}. The rate of dissociation is independent of the Ca^{2+} concentration. Ca^{2+} affects only the rate of association of actin with 1:1 gelsolin-actin complex .[7] The rate of dissociation of actin from the 1:2 gelsolin-actin complex turned out to depend on the source from which gelsolin was isolated. Actin dissociates from 1:2 complex in which gelsolin stems from human platelets, 50-fold more slowly than from the complex which contains chicken gelsolin.[7,8]

BINDING OF GELSOLIN TO POLYMERIC ACTIN

Association of gelsolin with actin filaments which is the first step of fragmentation of actin filaments by gelsolin, was followed by two experiments. First, the time course of this assembly reaction was measured by using a fluorescence label covalently bound to gelsolin. The fluorescence intensity of this label decreased, when gelsolin bound to actin filaments. According to a quantitative evaluation the fluorescently labeled gelsolin associates with actin filament subunits at a similar low rate as with actin monomers.[9] As there were indications that the fluorescence label attached to gelsolin affects the rate of association of gelsolin both with polymeric and with monomeric actin, a second experiment was designed in which unmodified gelsolin could be used. The rate of binding of gelsolin to actin filament subunits was measured by comparison with another competing reaction, namely binding of gelsolin to actin-DNase I complex. Gelsolin was added to a mixture of actin filaments and actin-DNase I complex. The fractions of gelsolin that bound to actin filament subunits or to actin-DNase I complex depends on the relative rates of these two competing reactions. Gelsolin that bound to actin filaments, was determined by the increase of the concentration of pointed filament ends created by fragmentation of actin filaments. Treadmilling actin filaments to which gelsolin was added, depolymerized at their newly formed pointed ends to reach the higher critical concentration of these ends because the barbed ends were capped by gelsolin (Fig. 4). In the presence of actin-DNase I

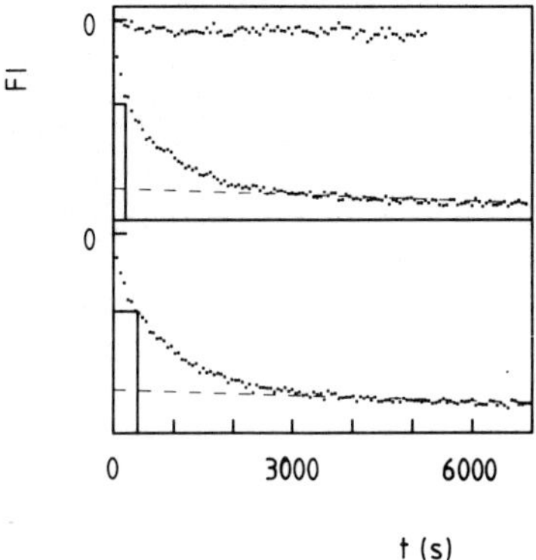

Fig. 4. Top panel, time course of depolymerization of 1 μM polymeric actin following addition of 30 nM gelsolin (lower curve). In a control experiment buffer was added to 1 μM polymeric actin (upper curve). Bottom panel, time course of depolymerization of 1 μM polymeric actin in the presence of 1 μM actin-DNase I complex following addition of 30 nM gelsolin. Depolymeristion was measured by the decrease of the fluorescence intensity FI (arbitrary units). The apparent half-life times are drawn in.

complex, depolymerization of the pointed ends following mixing with gelsolin was slower, because a fraction of gelsolin was consumed by binding to actin-DNase I complex and did not produce new pointed filament ends. It was found that, if equal concentrations of actin filament subunits and actin-DNase I complex were present, about half of the gelsolin bound to actin filament subunits and half to actin-DNase I complex (Fig. 4). This suggests that the two competing binding reactions occur at similar rates. The rate constant of association of gelsolin with actin-DNase I complex was then determined in the same way as association of gelsolin with actin monomers and turned out to be 1×10^4 M^{-1} s^{-1}. Thus, gelsolin binds also to actin filament subunits at a remarkably low rate of about 10^4 M^{-1} s^{-1}. Both the fluorescence and the competition measurements lead to the conclusion that gelsolin binds to the subunits of polymeric actin at a slow rate that is similar to the rate of formation of the 1:1 gelsolin-actin complex.[9]

CAPPING OF ACTIN FILAMENTS BY GELSOLIN-ACTIN COMPLEX

The equilibrium constant for binding of the gelsolin-actin complex to the barbed ends of actin filaments was measured by the depolymerizing effect of the gelsolin-actin complex on actin filaments. When the gelsolin-actin complex blocks monomer consumption at the polymerizing barbed ends of treadmilling actin filaments, actin monomers continue to be produced by depolymerization at the pointed ends until a new steady state is reached in which monomer production at the pointed ends is balanced by monomer consumption at the uncapped barbed ends. By using this effect the equilibrium constant K_{-Ca} for capping was roughly estimated to be about 10^{10} M^{-1} in excess EGTA over total calcium (Fig. 5). In the presence of micromolar Ca^{2+} concentrations the equilibrium constant of capping K_{+Ca} was estimated to be in the range of or above 10^{11} M^{-1}.

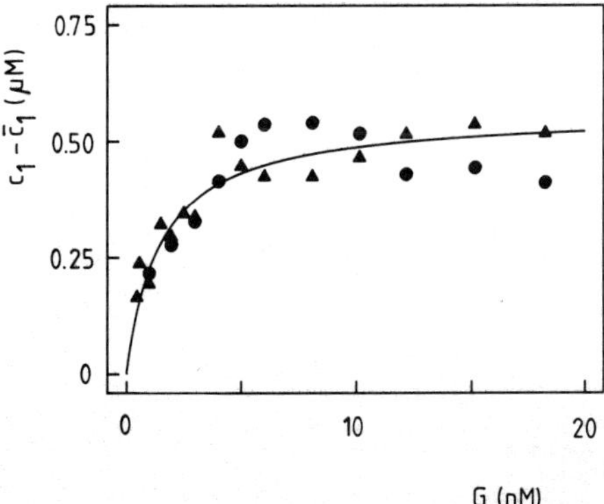

Fig. 5. Concentration of monomeric actin coexisting with polymeric actin in the presence of various concentrations of gelsolin-actin complex and in excess EGTA. c_1, concentration of actin monomers; $\bar{c}_1$, actin monomer concentration in the absence of gelsolin-actin complex when all barbed ends are free ($\approx .15$ μM). G, concentration of gelsolin-actin complex. The values were determined following depolymerization of polymeric actin (o) or polymerization of monomeric actin (▲). The continuous line was calculated for an equilibrium constant of capping $K_{-Ca} = 1.5 \times 10^{10}$ M^{-1}.

The rate of capping of barbed filament ends by the gelsolin-actin complex was measured by inhibition of polymerization of actin filaments (Fig. 6). Polymeric actin (0.1 - 1 μM) was added to 0.5 μM monomeric actin and various concentrations of gelsolin-actin complex (0.08 - 0.24 nM). Nucleation of new filaments by gelsolin-actin complex was avoided by keeping the monomer actin concentration below the critical concentration of the pointed ends (0.7 - 0.8 μM). The rate of binding of the gelsolin-actin complex to barbed filament ends was monitored by the time course of inhibition of polymerization. Gelsolin-actin complex was found to bind fourfold faster to the barbed ends in the presence of Ca^{2+} ($k_{+Ca}^{+} = 11 \times 10^6$ M^{-1} s^{-1}) than in excess EGTA ($k_{-Ca}^{+} = 2.5 \times 10^6$ M^{-1} s^{-1}).[10,11]

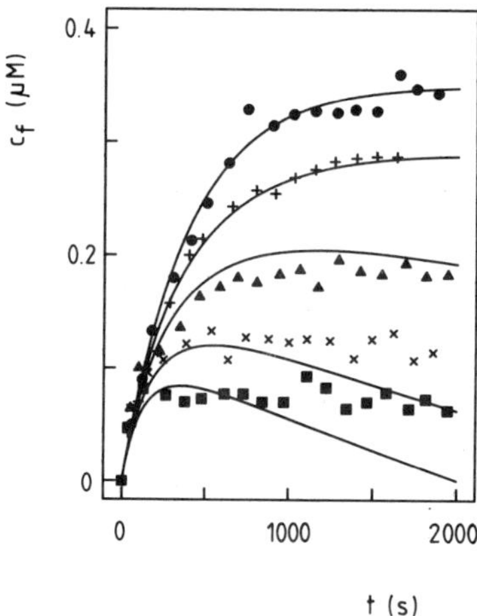

Fig. 6. Inhibition of actin polymerization by the gelsolin-actin complex. c_f, concentration of actin molecules incorporated into filaments. 0.5 μM monomeric actin was added to 0.4 μM polymeric actin in the presence of the following concentrations of gelsolin-actin complex: ■, 3 nM; x, 2 nM; ▲, 1 nM; +, 0.5 nM; o, 0 nM. The measurements were done in excess EGTA over Ca^{2+}. The continuous lines were calculated for a rate constant of capping $k_{-Ca}^{+} = 2.5 \times 10^6$ M^{-1} s^{-1}.

ADP-RIBOSYLATION OF GELSOLIN-ACTIN COMPLEXES

Various clostridial toxins such as *Clostridium botulinum* C2 toxin or *Clostridium perfringens* iota toxin ADP-ribosylate actin in arginine 177. The substrate for toxin-catalyzed ADP-ribosylation is monomeric actin but not polymeric actin.[12,13] The modified actin loses its ability to polymerize. Additionally, it has been shown that the ADP-ribosylated actin behaves like a capping protein which binds to the barbed ends of actin filaments to inhibit polymerization of unmodified actin at these ends.[14] Since gelsolin has been suggested to play a crucial role in regulation of the polymerization state of actin, we studied the functional consequences of the ADP-ribosylation of gelsolin actin complexes.[8] The actin molecules of the 1:1 and of the 1:2 gelsolin-actin complexes could be ADP-ribosylated by *Clostridium perfringens* iota toxin to form ADP-ribosylated 1:1 gelsolin-actin complex (G-A$_r$) or 1:2 gelsolin-actin complexes in which the first actin molecule (G-A$_r$-A), the second actin molecule (G-A-A$_r$) or both actin molecules (G-A$_r$-A$_r$) were ADP-ribosylated (A$_r$ denotes ADP-ribosylated actin). ADP-ribosylation did not affect the nucleation activity of the G-A$_r$ complex or the G-A$_r$-A complex. When monomeric actin was added to the G-A-A$_r$ complex, polymerization was delayed by about 10 min (Fig. 7). According to a quantitative kinetic analysis, the delay of polymerization corresponded to the rate of dissociation of ADP-ribosylated actin from the G-A-A$_r$ complex. This suggests that the nucleation activity of the G-A-A complex is inhibited by ADP-ribosylation of the weakly bound actin and that the inhibition can be removed by dissociation of ADP-ribosylated actin from the G-A-A$_r$ complex.

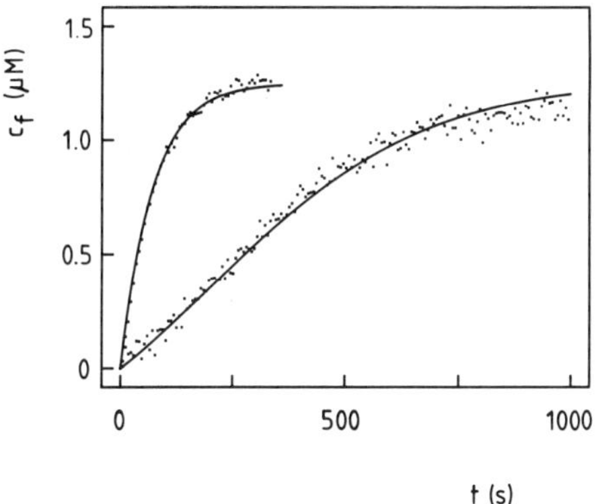

Fig. 7. Test for nucleation activity of the G-A-A$_r$ complex. c$_f$, concentration of polymerized actin. Dotted lower curve, 2 μM actin was added to 50 nM G-A-A$_r$ complex. Dotted upper curve, control experiment: 2 μM actin was added to 50 nM unmodified gelsolin-actin complex. The solid lower curve was calculated for the assumption that the G-A-A$_r$ complex is transformed into a nucleus by exchange of the ADP-ribosylated actin for unmodified actin. The solid upper curve was calculated for the assumption that the gelsolin-actin complex acts as a nucleus for actin polymerization.

ADP-ribosylated actin has been shown to behave like a capping protein which inhibits polymerization at the barbed ends but not at the pointed ends.[14] The finding that in the G-A-A$_r$ complex polymerization towards the pointed ends is inhibited by ADP-ribosylation of an actin molecule, suggests that the structural relationship of the two actin molecules in the 1:2 gelsolin-actin complex is different from the arrangement of the terminal subunits at the pointed end of an actin filament.

OPEN QUESTIONS

Careful investigations on the domain structure of gelsolin have revealed that within the six repeating sequence segments, domains S1 and S4-S6 bind to monomeric actin or to an actin filament subunit. Association with actin has been proposed to be prevented by internal interaction between S1 and S4-S6 in the absence of calcium. The two domains "open up" in calcium to render possible binding of gelsolin to actin.

Segments S2-S3 bind two adjacent filament subunits. Filaments are thought to be severed through binding of segments S2-S3 to adjacent actin filament subunits and disruption by segment S1.[15] The rates of all these intermediate steps remain an open question. In particular, it would be interesting to know the rate at which following binding of gelsolin to polymeric actin a filament is fragmented.

REFERENCES

1. H. L. Yin and T. P. Stossel, Control of Cytoplasmic Actin Gel-Sol Transformation by Gelsolin, a Calcium-Dependent Regulatory Protein, *Nature* 281:583 (1979).
2. R. Norberg, R. Thorstensson, G. Utter and A. Fagraeus, F-Actin-Depolymerizing Activity of Human Serum, *Eur. J. Biochem.* 100:575 (1979).
3. H. E. Harris and A. G. Weeds, Plasma Actin Depolymerizing Factor Has both Calcium-Dependent and Calcium-Independent Effects on Actin, *Biochemistry* 22:2728 (1983).
4. J. Bryan and M. C. Kurth, Actin-Gelsolin-Interactions: Evidence for Two Actin-Binding Sites, *J. Biol. Chem.* 259:7480 (1984).
5. P. A. Janmey and T. P. Stossel, Modulation of Gelsolin Function by Phosphatidyl 4,5-Bisphosphate, *Nature* 325:362 (1987).
6. N. Selve and A. Wegner, pH-dependent Rate of Formation of the Gelsolin-Actin Complex from Gelsolin and Monomeric Actin, *Eur. J. Biochem.* 168:111 (1987).
7. B. Schoepper and A. Wegner, Rate Constants and Equilibrium Constants for Binding of Actin to the 1:1 Gelsolin-Actin Complex, *Eur. J. Biochem.* 202:1127 (1991).
8. M. Wille, I. Just, A. Wegner, and K. Aktories, ADP-Ribosylation of Gelsolin-Actin Complexes by Clostridial Toxins, *J. Biol. Chem.* 267:50 (1992).
9. B. Schoepper and A. Wegner, Gelsolin Binds to Polymeric Actin at a Low Rate, *J. Biol. Chem.* 267:13924 (1992).
10. N. Selve and A. Wegner, Rate Constant for Capping of the Barbed Ends of Actin Filaments by the Gelsolin-Actin Complex, *Eur. J. Biochem.* 155:397 (1986).
11. N. Selve and A. Wegner, Rate Constants and Equilibrium Constants for Binding of the Gelsolin-Actin Complcx to the Barbed Ends of Actin Filaments in the Presence and Absence of Calcium, *Eur. J. Biochem.* 160:379 (1986).
12. K. Aktories, M. Bärmann, I. Ohishi, S. Tsuyama, K. H. Jakobs and E. Habermann, Botulinum C2 Toxin ADP-ribosylates Actin, *Nature* 322:390 (1986).
13. J. Vandekerckhove, B. Schering, M. Bärmann and K. Aktories, Botulinum C2 Toxin ADP-ribosylates Cytoplasmic ß/γ−Actin in Arginine 177, *J. Biol. Chem.* 263:696 (1988).
14. A. Wegner and K. Aktories, ADP-ribosylated Actin Caps the Barbed Ends of Actin Filaments, *J. Biol. Chem.* 263:13739 (1988).
15. B. Pope, M. Way and A. G. Weeds, Two of the Three Actin-Binding Domains of Gelsolin Bind to the Same Subdomaion of Actin: Implications for Capping and Severing Mechanisms, *FEBS Lett.* 280:70 (1991).

ACTIN REGULATION AND SURFACE CATALYSIS

Lawrence E. Crawford[1], Robert W. Tucker[2,3], Alan W. Heldman[1]
and Pascal J. Goldschmidt-Clermont[1,3]

1. Cardiology Division, Department of Medicine
2. Oncology Department
3. Department of Cell Biology and Anatomy
   Johns Hopkins University School of Medicine
   Baltimore, Maryland 21287

INTRODUCTION

The motile behavior of non-muscle cells often differs between healthy and pathological conditions. Two disease processes, cancer and atherosclerosis, are associated with high morbidity and mortality in our society. The cells involved in both the pathogenesis of and the defense against these diseases undergo marked changes in the organization of their actin cytoskeleton[1,2]. In response to a signal originating from the extracellular space, from surrounding cells, or as the result of a mutation, diseased cells initiate a process of motion away from their normal location. Local growth inhibitors are lost, and displaced cells undergo unchecked proliferation[1]. One example of such a phenomenon is the migration of fibroblasts and smooth muscle cells into the vascular intima and their proliferation in patients with atherosclerotic coronary artery disease[2]. Another example is the proliferation of metastatic cells distant from the site of primary tumor[1].

The migration of cells requires three essential steps. First, cells must detach from their normal environment[1]. Recent advances in the biology of proteases, like tissue plasminogen activator (TPA), and protease inhibitors such as plasminogen activator inhibitor 1 (PAI-1), have improved our understanding of this process. The disruption by proteases of specific interactions between integrins or other adhesive membrane proteins, and their extracellular ligand represents a limiting step in the ability of migrating cells to escape from their original location[1]. Second, cells must migrate to a new site. Migration represents a complex process involving both extracellular and intracellular molecules[3,4]. Regulation of the intracellular superstructure required to generate cell migration will be the focus of this review. Third, cells must establish new interactions with extracellular matrix proteins and other cells at the new site[1].

Two approaches have been used to study the molecules controlling cell migration. The first approach, biochemical, involves the characterization of the molecular constituents of the actin cytoskeleton and its membrane anchor[3,5]. The other approach, genetic, consists of studying genes whose disruption leads to abnormal motility of the affected cells[4]. Interestingly, the genetic approach has led to the identification of proteins which often do not interact directly with actin. Instead, membrane receptors or proteins secreted by the migrating cells and involved in their extracellular guidance, such as the product of *C Elegans* gene Unc-6 which is similar to the extracellular matrix protein laminin, have been identified as limiting for certain processes of migration[4]. It is likely that the redundancy of the proteins involved in the direct regulation of actin, or the lethal character of their deletion, may limit the

Actin: Biophysics, Biochemistry, and Cell Biology
Edited by J.E. Estes and P.J. Higgins, Plenum Press, New York, 1994

ability of the genetic approach to recognize these proteins as necessary for cell migration. Nevertheless, the concurrent use of biochemical and genetic approaches to study cell migration should lead to a clear understanding of the key steps involved in motility.

Here we will review three aspects of actin cytoskeleton control related to cell motility. First, we will review the role of membrane polyphosphoinositides such as phosphatidylinositol 4,5 trisphosphate (PIP2) as surface catalysts for specific surface reactions (reactions at the interface between cytoplasm and cell membrane) involving actin and actin binding proteins and leading to cytoskeletal rearrangement. Second, we will provide new insight about the regulation of key steps of the adenosine triphosphate (ATP) cycle in relationship to the dynamic control of actin structures. Third, we will summarize our preliminary data on the control provided by small guanine nucleotide (GTP) binding proteins for the organization of the actin cytoskeleton.

POLYPHOSPHOINOSITIDES AS SURFACE CATALYSTS FOR THE ACTIN CYTOSKELETON

Several actin binding proteins have been shown to interact with the head group of phosphatidylinositol (PI), phosphatidylinositol 4 phosphate (PIP), and PIP2[5,6]. Actually, whereas certain actin binding proteins bind to each of the three inositol phospholipids with similar affinity, other actin binding proteins, like gelsolin and profilin, bind only the polyphosphoinositides PIP and PIP2[6].

Over the past 15 years, substantial progress has been made in the chemistry of surface catalysis, the catalysis of reactions occurring at the interface between solid and liquid or gas phases. A direct application of this new branch of chemistry consists of the removal of carbon monoxide and nitric oxide from car exhaust fumes which is enhanced by surface catalysts[7]. By definition, a surface catalyst is a molecule located at the surface of a solid, to which soluble reactants bind transiently at the interface, thus facilitating a specific reaction by lowering the free energy associated with that reaction[7,8]. In the presence of surface catalysts, certain reactions will be favored and others will not, while the catalysts are unaffected by their participation in the catalyzed reactions[7,8].

Polyphosphoinositides, interacting with the proteins of the actin cytoskeleton, are, therefore, surface catalysts[6]. Their head groups, located at the interface between membrane and cytoplasm, interact with actin binding proteins to disrupt[6] or enhance[9] the interactions of these proteins with actin (Figure 1). The polyphosphoinositides are not affected directly by their interaction with actin binding proteins; there is no detectable hydrolysis or other modification of the polyphosphoinositides upon interaction with actin binding proteins[10]. Most important to their function in cell motility, the polyphosphoinositides are located in areas of cells where dynamic reorganization of the actin cytoskeleton occurs[11].

Membrane polyphosphoinositides are metabolized and translocated in response to surface receptor interaction with an extracellular ligand[12]. Therefore, the density of surface catalysts at specific membrane sites does vary with time and with cell stimulation. These changes are likely to induce corresponding regulation in the availability of actin binding proteins for interaction with actin (Figure 2).

As the polyphosphoinositides function as surface catalysts for the actin cytoskeleton, they are also substrates themselves for metabolism by phosphoinositide kinases and phospholipases[12]. PIP2 hydrolysis produces two second messengers, diacylglycerol (DAG) and inositol trisphosphate (IP3). Interestingly, actin binding proteins, upon binding to the polyphosphoinositides, can prevent their hydrolysis by certain phospholipase isoforms, an inhibition which can be overcome by activation of the phospholipase (e.g. by tyrosine phosphorylation) by membrane receptors[10].

Polyphosphoinositides represent a small proportion of membrane phospholipids. Their mixture with other lipids of various compositions does influence the surface catalytic activity of the polyphosphoinositides[10]. The specific mechanisms by which surrounding phospholipids influence interaction between actin binding proteins and polyphosphoinositides is not fully understood but may involve the ability of polyphosphoinositides to cluster[10,13,14]. Within synthetic membranes made of PIP2 and other phospholipids, five PIP2 head groups cluster to from one binding site for profilin[13,14]. In addition, agonist induced modification of the polyphosphoinositides, for example through PI3-kinase activity, which alters the hydrolysis of polyphosphoinositides by phospholipase

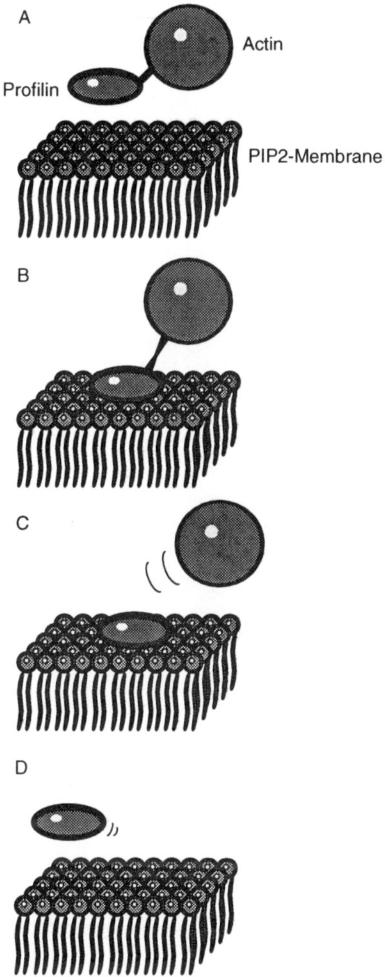

FIGURE 1. Polyphosphoinositides and Surface Catalysis for the Actin Superstructure.
Several actin binding proteins have been shown to interact with membrane polyphosphoinositides which function for the actin cytoskeleton as surface catalysts, facilitating reactions by lowering the energy required for the reactions to occur. The head groups of the polyphosphoinositides can cluster at the interface between membrane and cytoplasm to form binding sites for the actin binding proteins. Binding to these clusters destabilizes the interaction between the regulatory proteins and actin. Interaction between the polyphosphoinositides and actin binding proteins is usually transient and therefore, the actin binding proteins cycle back to the cytoplasm and are free to interact with other actin molecules. Although the polyphosphoinositides are not metabolized as a result of interaction with the actin binding proteins, they represent substrates for specific kinase, phosphatases, and phospholipases and their availability for these enzymes can be altered by interaction of with the actin binding proteins.

C, may also modify the interaction of the polyphosphoinositides with the actin binding proteins[12].

ACTIN NUCLEOTIDE CYCLE AND FILAMENT DYNAMICS

Actin is a nucleotide triphosphatase which oscillates between several energy states in a fashion which has been recently characterized [Tirion MM, *et al.*: Actin '92-Meeting (abstract)]. Actin's interaction with its adenine nucleotide (ATP vs ADP) plays an important role in controlling actin structure. The sustained dissociation of actin from its nucleotide induces the stabilization of a monomeric actin state prohibiting self-assembly into filaments. Actin monomers at that stage are denatured as their conformation change cannot be reversed. Actin monomers bound to ATP polymerize more readily because they generate filaments with a high affinity end for the surrounding actin subunits in the cytoplasm[16]. The high affinity, barbed end, of actin filaments are usually not found when filaments are made of ADP actin monomers[16] (Figure 3).

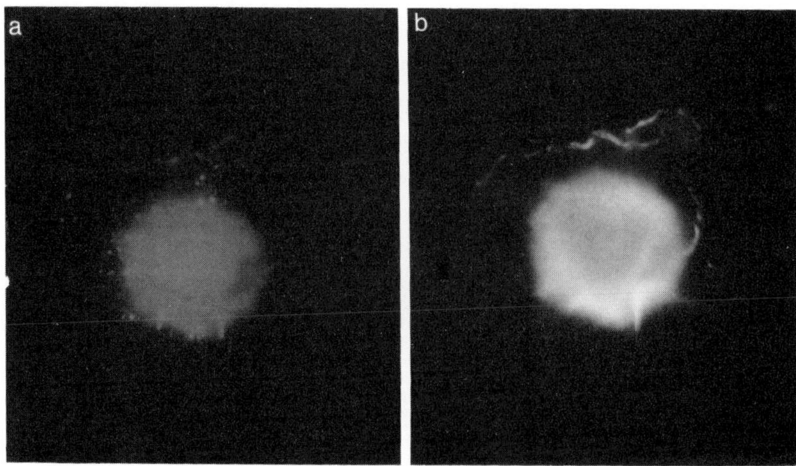

Figure 2. Comparison of Profilin and PIP2 Distribution in Chinese Hamster Ovary Cells.
Chinese hamster ovary cells (CHO-cells) were double-stained for profilin (panel a) and PIP2 (panel b). Formaldehyde-fixed CHO-cells were permeabilized using a quick-dip in acetone (-20°C) then stained with a rabbit antiserum against mammalian profilin (rhodamine-labeled goat anti-rabbit secondary antibody), and also a monoclonal antibody against PIP2 (FITC-labeled goat anti-mouse secondary antibody). Although the distributions of profilin and PIP2 seem to overlap in many area of the CHO-cells, they are not strictly identical. In particular, the concentration of profilin within several "spots" within the perinuclear area, where actin monomers have been shown to cluster, is not matched by PIP2 molecules.

There is some controversy with regard to the stability of filaments made of ATP versus ADP actin monomers[17,18]. It is likely that the methods used for the purification of actin and its preservation play a role in the generation of inconsistent data between laboratories with regard to the stability of actin filaments polymerized from ATP versus ADP actin subunits. It is possible that in non-muscle cells where the turnover of actin filaments is remarkably fast[19,20], actin filaments made of ADP actin subunits might be particularly unstable. The efficiency of severing proteins (such as gelsolin), may be affected by the nucleotide bound to actin monomers, ATP versus ADP, within the milieu surrounding these proteins (Janmey PA, personal communication).

Most actin binding proteins, such as for example, gelsolin, cofilin, thymosin beta4 and DNAse-1, have been shown to inhibit the exchange of the nucleotide bound to actin. In contrast, profilin is the only actin binding protein which has been shown to increase the rate of exchange of nucleotide bound to actin[21,22].

ATP nucleotide hydrolysis is also affected by actin binding proteins. Gelsolin (Janmey PA, personal communication), as well as other actin subunits within actin polymers, have been shown to increase the rate of nucleotide hydrolysis by actin[23]. Profilin, conversely, has been shown to reduce actin triphosphatase activity[21]. In addition, toxins of the cytochalasin family, known to disrupt the actin cytoskeleton in cells are very strong activators of actin monomer triphosphatase activity in the presence of magnesium[24].

The careful characterization of the effect of additional actin binding proteins on the interaction of actin with its nucleotide will be necessary to understand the details of the regulation of the actin cytoskeleton in cells. Actually, this new approach to the interaction of actin with its satellite proteins may allow us to simplify the classification of the actin binding

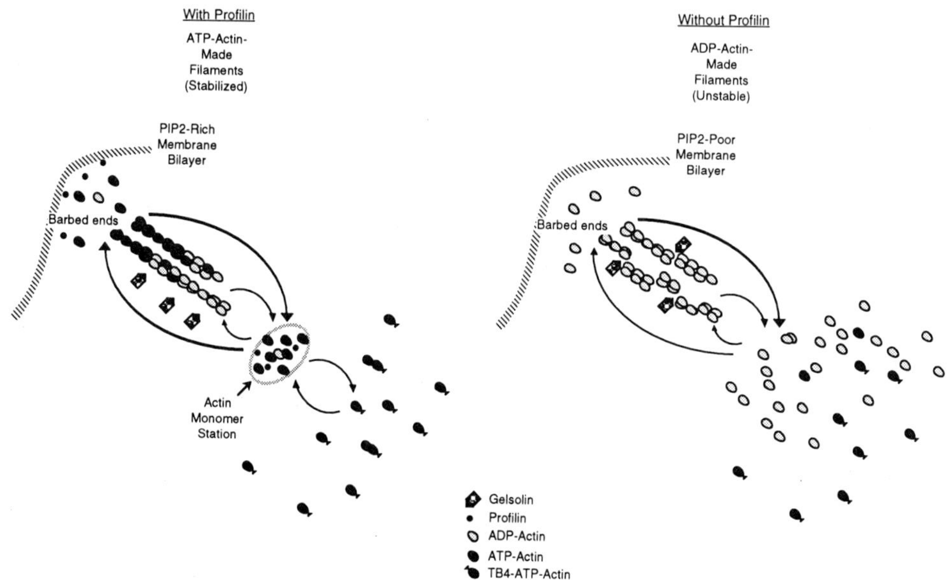

FIGURE 3. The Regulation of Actin Monomers.
Actin is a nucleotide triphosphatase regulated by proteins controlling the interaction of monomeric actin with its bound adenine nucleotide. While most actin binding proteins seem to inhibit the exchange of the nucleotide bound to actin, profilin accelerates this nucleotide exchange. Profilin's effect competes with the opposite effect of the most abundant actin monomer binding protein in many cells, thymosin β4. Carlier et al have recently shown that thymosin β4 binds with much higher affinity to ATP actin monomers than to ADP actin monomers (~ 50 fold difference). In cells where actin filaments turn-over rapidly and large amounts of ADP actin monomers are released from the ends of filaments, our model predicts that, in the absence of profilin, actin polymers will be made mainly of ADP actin monomers, as actin monomers recharged with ATP are selectively sequestered by thymosin β4. However, in the area of cells where profilin is abundant, ADP would be exchanged so rapidly for ATP on monomers that actin polymers would be made primarily of ATP actin monomers. Although, the destabilizing effect of ADP monomers for actin filaments made in vitro is still controversial, in cells, actin filaments made of ATP actin monomer seems to be more stable than filaments made of ADP actin subunits.

proteins. The effect of actin binding proteins on the actin-bound nucleotide may represent a reliable predictor of those proteins effect on actin inside cells.

SMALL GTP BINDING PROTEINS AND REGULATION OF ACTIN CHANGES

Transient changes in the actin cytoskeleton can be induced by growth factor interaction with a membrane receptor[25,26]. The resulting cyclic and orchestrated changes occurring in migrating cells are likely to require further mechanisms of regulation. Small GTP binding proteins (G-proteins) are poised to fulfill the function of regulators for the assembly and disassembly of complex actin structures[27,28]. First, they do regulate in other systems the assembly of DNA, protein, and nucleotide complexes necessary for DNA transcription. Second, transformation of cells with a constitutively activated form of the G-protein Ras has been shown to induce marked changes in the cytoskeletal morphology of the transformed cells. Third, two proteins of the Ras family, Rho and Rac, have been shown to regulate the stability of focal adhesions and the formation of membrane ruffles, respectively[25,26,29].

We have initiated studies aimed at characterizing the mechanism by which a transforming isoform of Ras (Ha-Ras) can induce the reorganization of the actin skeleton. We observed that Ha-Ras transformed NIH-3T3 fibroblasts contain an actin superstructure which is organized in a fashion similar to that of non-transformed cells activated with growth factors[25,26] (Figure 4). Most of the filamentous (F-) actin in the Ha-Ras transformed cells is found in the cortical area of the cell and is particularly abundant in area of cell extensions such as ruffling. In addition, the abundant stress fibers which are found in fibroblasts exposed to low concentration of serum (0.5%) are practically absent from Ras-transformed fibroblasts. When we analyzed cell lysates from NIH-3T3 cells and from the same cells transformed with Ha-Ras, we find that the total actin concentration is not altered by Ha-Ras when normalized for total protein content. However, the fraction of actin present in the filamentous form was reduced in the Ha-Ras transformed cells.

Increased ruffling, concentration of actin filaments in the cortical area of the cell and the paucity of stress fibers induced by Ha-Ras represent features usually present early in the activation of cells by growth factors such as platelet-derived growth factor (PDGF)[25,26]. We next performed experiments to test whether PDGF could still affect motility and actin cytoskeletal organization in cells transformed with Ha-Ras. In both NIH-3T3 cells and Ha-Ras transformed cells, PDGF induced a rapid (within one minute) increase in membrane ruffling, associated with a shift of actin subunits from the monomeric to the filamentous pool. We also observed that the tyrosine kinase signalling pathway of the PDGF receptor is essentially intact: on agonist binding, the PDGF receptor phosphorylates itself on tyrosine, and also phosphorylates phospholipase Cγ1 (PLCγ1). However, the time course of these phosphorylation reactions is altered by Ha-Ras transformation. In addition, despite the activation of PLCγ1, the calcium transient normally observed in response to the burst of IP3 produced by the activated phospholipase C is absent in Ha-Ras transformed cells. The lack of calcium transient is not explained by absence of IP3 receptor, as Ha-Ras transformed cells contain at least as much IP3 receptor as do the parent cells (Heldman et al., unpublished results). Instead, the absence of IP3-burst may be responsible for the lacking calcium transient.

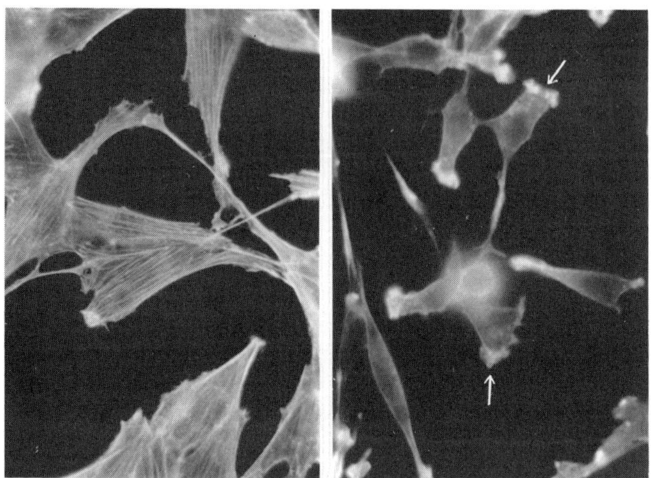

FIGURE 4. Effect of Ha-Ras Transformation on Actin Superstructure
NIH 3T3 cells transfected with the neomycin resistance gene (left panel) or stably expressing transforming Ha-Ras (right panel) were grown in DMEM supplemented with 0.5% serum, 1.0% penicillin/streptomycin and G418 (0.25 mg/ml). The cells were incubated for 48 hours in this medium prior fixation in formaldehyde (3.7% in PBS). Cells were permeabilized in Triton X 100 (0.2% in PBS), and were stained for filamentous actin with Rhodamine Phalloidin (0.2 micromolar in PBS). NIH 3T3 cells display the well-characterized actin stress fibers anchored at the membrane level to the substratum by focal adhesions. Ha-Ras transformed cells contain a very different actin superstructure. Actin stress fibers are practically undetectable, and most filamentous actin is localized underneath the plasma membrane, particularly in area of cellular extensions. The intense membrane ruffling activity can also be appreciated on this micrograph (white arrows), a feature which is clearly associated with the Ras transformed phenotype, even in the presence of "starving" concentrations of serum. Another typical feature of Ras transformed cells is the retracted and fusiform shape of the cells, with ruffles that can extend several microns beyond the roof of the cell body.

In summary, Ha-Ras transformation of NIH-3T3 cells constitutively locks these cells in the activated state and prevents them from switching back to G_0 phase, but does not abolish the actin response to growth factor. The absence of calcium transient in response to growth factor despite the preserved actin response indicates that the calcium transient is unlikely to be responsible for the regulation of the actin response. The characterization of the molecular reactions linking the small GTP binding proteins and the actin cytoskeleton are underway. Careful analysis of the movement and metabolism of the polyphosphoinositides, relative to actin satellite proteins, should provide valuable information regarding cellular actin, and the regulation of its structural organization in resting cells, and cells triggered to become motile.

REFERENCES

1. Van Roy F, Mareel M, Tumor invasion: effects of cell adhesion and motility. *Trends Cell Biol* 2:163 (1992).
2. Heldman AW, Furman MI, Gardner TM, Gips SJ, Crawford LE, Goldschmidt-Clermont PJ, Coronary artery disease and atherogenesis. *in*: Molecular Basis of Medicine, Dang CV and Feldman AM, eds., (1993) in press.
3. Cooper JA, The role of actin polymerization in cell motility. *Annu Rev Physiol* 53:585 (1991).
4. Howard K, Getting there? *Curr Biol* 3:103 (1993).
5. Luna EJ, Hitt AL, Cytoskeleton-plasma membrane interactions. *Science* 258:955 (1992).
6. Goldschmidt-Clermont PJ, Janmey PA, Profilin, a weak CAP for actin and RAS. *Cell* 66,419 (1991).
7. Friend CM, Catalysis on surfaces. *Scientific American* 268:74 (1993).
8. Yates JT, Surface chemistry. *Chemical Engineering News* 70:22 (1992).
9. Fukami K, Furuhashi K, Inagaki M, Endo T, Hatano S, Takenawa T, Requirement of phosphatidylinositol 4,5 bisphosphate for α–actinin function. *Nature* 359:150 (1992).
10. Goldschmidt-Clermont PJ, Kim JW, Machesky LM, Rhee SG, Pollard TD, Regulation of phospholipase Cγ1 by profilin and tyrosine phosphorylation. *Science* 251:1231 (1991).
11. Finkel T, Theriot JA, Dise KR, Tomaselli GF, Goldschmidt-Clermont PJ, Actin superstructure promoted by profilin. *Proc Natl Acad Sci USA* (1993) in press.
12. Cantley LC, Auger KR, Carpenter C, Duckworth B, Graziani A, Kapeller R, Soltoff S, Oncogenes and signal transduction. *Cell* 64:281 (1991).
13. Goldschmidt-Clermont PJ, Machesky LM, Baldassare JJ, Pollard TD, The actin-binding protein profilin binds to PIP2 and inhibits its hydrolysis by phospholipase-C. *Science* 247:1575 (1990).
14. Machesky LM, Goldschmidt-Clermont PJ, Pollard TD, The affinities of human platelet and *Acanthamoeba* profilin isoforms for polyphosphoinositides account for their relative abilities to inhibit phospholipase C. *Cell Regul* 1:937-950 (1990).
15. Engel J, Fasold H, Hulla FW, Waechter F, Wegner A, The polymerization reaction of muscle actin. *Mol Cell Biochem* 18:3 (1977).
16. Wegner A, Head to tail polymerization of actin. *J Mol Biol* 108:139-150, 1976.
17. Janmey PA, Hvidt S, Oster GF, Lamb J, Stossel TP, Hartwig JH, Effect of ATP on actin filament stiffness. *Nature* 347:95 (1990).
18. Pollard TD, Goldberg I, Schwarz WH, Nucleotide exchange, structure, and mechanical properties of filaments assembled from ATP-actin and ADP-actin. *J Biol Chem* 267:20339 (1992).
19. Theriot JA, Mitchison TJ, Actin microfilaments dynamics in locomoting cells. *Nature* 352:126 (1991).
20. Theriot JA, Mitchison TJ, The nucleation-release model of actin filament dynamics in cell motility. *Trends Cell Biol.* 2:219 (1992).
21. Mockrin SC, Korn ED, *Acanthamoeba* profilin interacts with G-actin to increase the rate of exchange of actin-bound adenosine 5'-triphosphate. *Biochemistry* 19:5359 (1980).
22. Goldschmidt-Clermont PJ, Furman MI, Wachsstock D, Safer D, Nachmias VT, Pollard TD, The control of actin nucleotide exchange by thymosinb4 and profilin. A potential regulatory mechanism for actin polymerization in cells. *Mol Biol Cell* 3:1015 (1992).
23. Carlier M-F, Role of nucleotide hydrolysis in the dynamics of actin filaments and microtubules. *Int Rev Cytol* 115:139 (1989).
24. Cooper JA, Effects of cytochalasin and phalloidin on actin. *J Cell Biol* 105:1473 (1987).
25. Ridley AJ, Hall A, The small GTP-binding protein rho regulates the assembly of focal adhesions and actin stress fibers in response to growth factors. *Cell* 70:389 (1992).
26. Ridley AJ, Paterson HF, Johnston CL, Diekmann D, Hall A, The small GTP-binding protein rac regulates growth factor-induced membrane ruffling. *Cell* 70:401 (1992).
27. Bourne HR, Sanders DA, McCormick F, The GTPase superfamily: a conserved switch for diverse cell functions. *Nature* 348:125 (1990).
28. Bourne HR, Sanders DA, McCormick F, The GTPase superfamily: conserved structure and molecular mechanism. *Nature* 349:117 (1991).

29.Goldschmidt-Clermont PJ, Mendelsohn ME, Gibbs JB, Rac and Rho in control. *Curr Biol* 2:669 (1992).
30.Acknowledgements: This research was supported in part by a grant from Syntex, by a grant from the Bernard Foundation and by the American heart Association (G-I-A, Maryland Affiliate, Inc.). PJG-C was selected as a Syntex Scholar in 1992.

CALDESMON: POSSIBLE FUNCTIONS IN MICROFILAMENT RE-ORGANIZATION DURING MITOSIS AND CELL TRANSFORMATION

Shigeko Yamashiro, Kyonsoo Yoshida, Yoshihiko Yamakita, and
Fumio Matsumura

Department of Molecular Biology and Biochemistry
Nelson Labs/busch Campus
Rutgers University
Piscataway, NJ 08855-1059

INTRODUCTION

Microfilaments are intimately involved in the morphological alterations observed both in cell transformation and mitosis. In both cases, microfilamenets show significant re-organization. While "normal" cells with well-spread morphologies have numerous bundles of microfilaments, transformed cells with rounded morphologies show dispersed microfilament patterns[1]. These changes in the structure of actin cables, as well as in substrate adhesion, are closely coupled with proliferation in both normal and transformed cells[1], suggesting that the microfilament cytoskeleton may play an important role in oncogenic transformation. The process of cell division causes similar alterations in microfilament patterns to those seen in transformed cells. Microfilament bundles disassemble when cells become rounded-up during prophase[2,3] resulting in a cytoarchitecture resembling the dispersed microfilament patterns seen in the many transformed cell types which exhibit rounded morphologies. Microfilament bundles that are anchored to focal contacts are also disrupted in both mitotic and transformed cells, causing reduced adhesion of such cells to their substrates. These observations suggest that cell transformation and cell division control are intimately related, not just superficially, but at the level of molecular control.

Despite this close relationship between cell transformation and cell division, our biochemical understanding of the molecular mechanisms responsible for the alterations seen in each of these processes remains incomplete. Little is known regarding how the alterations in microfilament patterns seen in transformation and mitosis are related to each other at the molecular level.

Caldesmon, one of microfilament-associated proteins, may play a key role in microfilament assembly in both cell transformation and cell division. We have observed that caldesmon is disassociated from microfilaments as a result of mitosis-specific phosphorylation[4]. This finding is complimented by the observation that expression of caldesmon is decreased upon cell transformation[5,6]. These findings have functional relevance. Caldesmon is known to inhibit actomyosin ATPase[7], and caldesmon and tropomyosin together regulate the actin severing and capping activities of gelsolin[8,9]. In addition, caldesmon is reported to stimulate actin polymerization. All of these functions are clearly involved in the regulation of cell shape. In this report we first briefly review the *in vitro* functions of caldesmon. We will then describe our studies including some new data, which have contributed to the establishment of caldesmon as an important regulatory protein. Finnaly,we will discuss possible functions of caldesmon in the re-organization of microfilaments observed in these two biologically important phenomena.

IN VITRO FUNCTIONS OF CALDESMON

Caldesmon is a regulatory protein implicated in the control of actomyosin interactions in smooth muscle and nonmuscle cells (see ref[7] for review). This protein, first isolated from smooth muscle, has actin-binding and calmodulin-binding properties[10]. Two major isotypes of caldesmon have been identified: a high molecular weight form in smooth muscle estimated from SDS gel electrophoresis at $Mr = 120,000-150,000$; and a lower molecular weight species found mainly in nonmuscle cells estimated at $Mr = 70,000-83,000$[6,11]. Smooth muscle and nonmuscle caldesmons are localized on thin filaments and stress fibers, respectively, suggesting their involvement in the regulation of cell motility and/or microfilament organization.

Both muscle and nonmuscle caldesmons have similar properties, including a rod-like molecular shape, heat-stability, calmodulin regulated actin binding, self-association through disulfide bonds, periodic localization along actin filaments, stimulation of the actin binding properties of tropomyosin, and incorporation into stress fibers in fibroblasts[6,11-16]. This similarity is well explained by recent reports, which have shown through characterization of cloned cDNAs that the two caldesmons are highly homologous. Nonmuscle caldesmon is smaller, lacking 232 amino acids found in the central region of smooth muscle caldesmon[17,18]. The sequence data, together with studies on domain mapping by others[6,12,19,20], have placed both the actin- and calmodulin-binding domains at the C-terminal end of the caldesmon molecule and the myosin binding domain at the N-terminus.

Caldesmon has three major *in vitro* functions. First, caldesmon inhibits the tropomyosin-stimulated, actin-activated ATPase of myosin in reconstituted systems[21-26]. This inhibition is attenuated by Ca^{2+}/calmodulin, which reverses actin binding by caldesmon. Second, caldesmon inhibits actin capping and severing activities of gelsolin[8,9]. Third, caldesmon is reported to stimulate actin polymerization[27,28].

CALDESMON IN MITOSIS

Our understanding of the profound changes in cell shape which occur during mitosis is very limited. We do know that there are dramatic changes in the organization of microfilaments[2,3]. These changes include the disassembly of microfilament bundles during prophase which is accompanied by the rounding-up of cultured cells, the formation of transient contractile rings during cytokinesis, and, subsequently, the reassembly of microfilament bundles and the respreading of the daughter cells. In addition, it has been demonstrated by microinjection of antibodies against myosin II, as well as by disruption of the myosin II gene, that myosin II is involved in cytokinesis[29-31], though its mechanism of action remains unclear.

We have found that caldesmon is dissociated from microfilaments during mitosis, apparently as a consequence of mitosis-specific phosphorylation[32]. We have further demonstrated that p34[cdc2] kinase is responsible for the mitosis-specific phosphorylation of caldesmon. The finding that caldesmon is phosphorylated by cdc2 kinase strongly suggests that cdc2 kinase *directly* controls the changes in microfilament assembly seen during mitosis. Other examples of direct action of cdc2 kinase on cellular structures have been recently reported. The phosphorylation of both the intermediate filament protein vimentin and nuclear lamin by cdc2 kinase has been shown to induce the disassembly of either intermediate filaments or the nuclear lamina[33,34]. Because microfilament rearrangement, as well as the disassembly of intermediate filaments and the nuclear lamina are all observed in early stages of mitosis, the direct involvement of cdc2 kinase in these events appears reasonable.

To explore the effects of the mitosis-specific phosphorylation of caldesmon, *in vivo* and *in vitro* phosphorylated caldesmons have been characterized. We have found that both *in vivo* and *in vitro* phosphorylation of caldesmon causes the loss of most of caldesmon's properties, including its inhibition of actomyosin ATPase, and its actin- calmodulin- and myosin-binding properties. Rat nonmuscle caldesmon is phosphorylated *in vitro* up to a ratio of 7 moles of phosphate per mol of protein. Actin-bound caldesmon can be phosphorylated by cdc2 kinase, resulting in dissociation from F-actin. These observations suggest that the inhibition of most of caldesmon's functions may be a required step in the massive re-organization of microfilaments seen during mitosis.

Mitosis-specific phosphorylation of caldesmon affects the actin binding of other actin binding protein in reconstituted systems. For example, the dissociation of caldesmon by mitosis-specific phosphorylation reduces actin binding of tropomyosin as actin binding

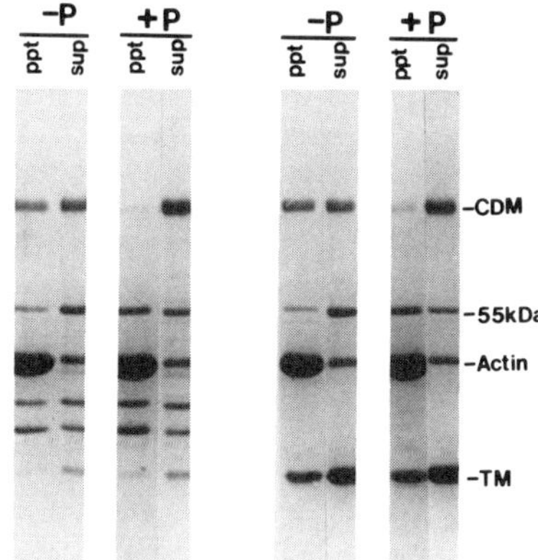

Figure 1. Dissociation of caldesmon by mitosis-specific phosphorylation affects actin binding of high and low Mr tropomyosin isoforms. -P, unphosphorylated caldesmon; +P, phosphorylated caldesmon; CDM. caldesmon; TM, tropomyosin. Note that, when caldesmon is phosphorylated, caldesmon is dissociated from actin, and actin binidng of 55kDa protein is increased. On the other hand, when caldesmon is unphosphorylated, caldesmon binds to actin, and the binding of 55kDa protein is inhibited. Similar results are obtaincd with either high Mr or low Mr tropomyosin isoforms.

of tropomyosin is stimulated by actin binding of caldesmon. Another example is the effects of caldesmon on actin binding of HeLa 55kDa actin bundling protein (recently identified as a human homologue of fascin). 55kDa actin-bundling protein inhibits the actin binding properties of the tropomyosins[35]. When caldesmon is present, 55kDa protein can no longer dissociate tropomyosin from actin. Instead, actin binding of HeLa 55kDa actin binding protein is inhibited by caldesmon together with tropomyosin. This inhibition is released when caldesmon is dissociated from actin filaments by phosphorylation with cdc2 kinase (Fig. 1). These results suggest that caldesmon has a major role in regulating the binding of tropomyosin to actin, and that the dissociation of cadesmon could greatly affect the microfilament organization during mitosis .

Caldesmon appears to be dissociated from microfilaments not only during mitosis but also during cytokinesis. We have examined the relationships between the phosphorylation level, actin-binding, and *in vivo* localization of caldesmon in cultured cells after the release from metaphase arrest[36]. Immunofluorescence studies have revealed that

    1 CTGCAAGAGCAAGATATTTGTGCGGAATGATCATGGGTGGGGCCAAGAGAAATTTAAGAAAGCTTTTAATTCTCGATGTGTTGTCCTGGG
   91 TAACTAGGGAAGATGCTGGAGAGACCAGGATTGCACTTTTCTGTGACAGGGGATGGAAATCAGATCGCAGAATTTGTCTCAGGATGCTCC
  181 ATGGACAGGAAATGGTGATGCTAGAGTCCTGGTGAGAAGTCTCCACTCAGCACTGGAGGCTCAGGCCTTTCTTCATCCTCGGGCCTGGAG
  271 ACAGCGATCAGTCTTGGTTAGATCAATGATATTGTCCCTAGTGATTCCATGCGCCTCACATAGTTTTGCAAACTTTTCCTTGATGTCTGA
  361 ACTCAGATCCTTTGTTCTGCCGTAGAGCACCATCAGCTGGAAGGTTTCCCCGTTCTTGAAATTAATGAGATGAAACATGACATATCTGTC
  451 ATAGTCTGTCTTAAGTATAGTAAATGTATTCCCTCCGTCATACTCAACAAAATATTCGGAATTCCGACTGGCTTGTTTTATTGAGGGGGG
  541 ACATGGGTAAGAACAAGTAAGACAGTTGAGACCCAAGTGTTAACACAAAGGGAAGACTGGCGCATCCTGCTCAGCGCATTTCGGCAACAG
  631 ATAATATGGCCCGTGTGCTGGCCCAGCCCATGCTCTGACCTCCAGGCGCCAGGTCCCCTACCTCAGTCACTCCTACCCAGACCAGCTGCG
  721 GACATGCTTAGCAGATCCGGGTCCCAGGGAAGGCGCTGCCTGGCCACACTCTCTCAAATTGCCTATCAGAGGAATGATGATGACGAAGAA
          M   L   S   R   S   G   S   Q   G   R   R   C   L   A   T   L   S   Q   I   A   Y   Q   R   N   D   D   D   E   E   29
  811 GAGGCTGCCAGGGAACGGCGCCGCCGAGCCCGACAGGAAAGGCTGCGGCAGAAGCAAGAGGAAGAATCCTTGGGACAGGTGACAGACCAA
          E   A   A   R   E   R   R   R   R   A   R   Q   E   R   L   R   Q   K   Q   E   E   E   S   L   G   Q   V   T   D   Q   59
  901 GTGGAGGCCCATGTCCAGAACAGTGCCCCCGATGAAGAGTCTAAGCCAGCCACTGCAAATGCTCAGGTAGAAGGTGATGAGGAGGCTGCT
          V   E   A   H   V   Q   N   S   A   P   D   E   E   S   K   P   A   T   A   N   A   Q   V   E   G   D   E   E   A   A   89
  991 TTGCTGGAGCGCCTGGCCAGGCGAGAAGAGAGACGTCAAAAACGCCTTCAGGAAGCCTTAGAGCGTCAGAAGGAGTTTGATCCAACCATA
          L   L   E   R   L   A   R   R   E   E   R   R   Q   K   R   L   Q   E   A   L   E   R   Q   K   E   F   D   P   T   I   119
 1081 ACCGATGGCAGTCTCTCAGTCCCAAGCAGAAGGATGCAAAATAACTCGGCTGAAAATGAGACAGCAGAGGGGGAAGAAAAAGGAGAGAGT
          T   D   G   S   L   S   V   P   S   R   R   M   Q   N   N   S   A   E   N   E   T   A   E   G   E   E   K   G   E   S   149
 1171 CGCTCAGGACGGTATGAGATGGAAGAAACAGAAGTGGTCATCACGTCCTACCAGAAGAACAGCTATCAGGATGCTGAAGACAAAAAGAAA
          R   S   G   R   Y   E   M   E   E   T   E   V   V   I   T   S   Y   Q   K   N   S   Y   Q   D   A   E   D   K   K   K   179
 1261 GAAGAAAAGGAGGAGGAGGAGGAGGAAGAGAAGCTGAAGGGAGGGAACCTTGGCGAAAATCAGATCAAAGATGAGAAGATTAAAAAGGAC
          E   E   K   E   E   E   E   E   E   E   K   L   K   G   G   N   L   G   E   N   Q   I   K   D   E   K   I   K   K   D   209
 1351 AAAGAGCCCAAAGAAGAAGTCAAGAACTTCTTGGATCGAAAGAAAGGATTTACAGAAGTGAAGGCGCAGAATGGAGAATTCATGACCCAC
          K   E   P   K   E   E   V   K   N   F   L   D   R   K   K   G   F   T   E   V   K   A   Q   N   G   E   F   M   T   H   239
 1441 AAACTTAAACAAACTGAGAATGCTTTCAGCCCCAGCCGTTCAGGAGGCCGGGCCAGCGGGGACAAGGAAGCTGAAGGCGCCCCACAAGTG
          K   L   K   Q   T   E   N   A   F   S   P   S   R   S   G   G   R   A   S   G   D   K   E   A   E   G   A   P   Q   V   269
 1531 GAAGCCGGTAAGAGGCTGGAGGAGCTGCGCCGACGCCGTGGGGAGACAGAGAGCGAAGAGTTCGAAAAACTCAAACAAAAGCAACAGGAG
          E   A   G   K   R   L   E   E   L   R   R   R   R   G   E   T   E   S   E   E   F   E   K   L   K   Q   K   Q   Q   E   299
 1621 GCAGCCTTGGAGCTGGAAGAGCTGAAGAAAAAGAGGGAAGAGAGAAGGAAGGTTCTGGAGGAAGAGGAGCAGAGAAGGAAGCAGGAGGAG
          A   A   L   E   L   E   E   L   K   K   K   R   E   E   R   R   K   V   L   E   E   E   Q   R   R   K   Q   E   E   329
 1711 GCTGATCGAAAAGCTAGAGAGGAGGAAGAGAAGAGGAGGTTGAAGGAAGAGATCGAAAGGAGAAGGGCAGAAGCTGCTGAGAAACGCCAG
          A   D   R   K   A   R   E   E   E   E   K   R   R   L   K   E   E   I   E   R   R   R   A   E   A   A   E   K   R   Q   359
 1801 AAGATGCCGGAAGATGGCCTATCTGAGGACAAGAAGCCGTTCAAGTGCTTCACTCCTAAAGGCTCATCTCTCAAGATAGAGGAGCGAGCA
          K   M   P   E   D   G   L   S   E   D   K   K   P   F   K   C   F   T   P   K   G   S   S   L   K   I   E   E   R   A   389
 1891 GAGTTTTTGAATAAGTCTGTGCAGAAAAGTGGTGTCAAATCAACTCATCAAGCAGCTGTGGTCTCCAAGATTGACAGCCGGCTGGAGCAA
          E   F   L   N   K   S   V   Q   K   S   G   V   K   S   T   H   Q   A   A   V   V   S   K   I   D   S   R   L   E   Q   419
 1981 TATACCAATGCAATCGAGGGAACAAAAGCTTCAAAACCTATGAAGCCTGCAGCATCCGATCTTCCTGTCCCTGCGGAAGGTGTCCGCAAT
          Y   T   N   A   I   E   G   T   K   A   S   K   P   M   K   P   A   A   S   D   L   P   V   P   A   E   G   V   R   N   449
 2071 ATCAAGAGCATGTGGGAGAAAGGGAGTGTGTTTTCATCCCCCTCTGCCTCGGGGACACCAAATAAGGAAACTGCTGGCCTGAAGGTGGGG
          I   K   S   M   W   E   K   G   S   V   F   S   S   P   S   A   S   G   T   P   N   K   E   T   A   G   L   K   V   G   479
 2161 GTTTCCAGCCGCATCAATGAATGGCTAACTAAATCGCCGGACGGCAACAAGTCACCCGCTCCCAAGCCTTCTGACTTAAGGCCAGGAGAT
          V   S   S   R   I   N   E   W   L   T   K   S   P   D   G   N   K   S   P   A   P   K   P   S   D   L   R   P   G   D   509
 2251 GTATCTGGCAAGCGGAACCTCTGGGAAAAGCAATCCGTGGATAAGGTCACTTCTCCCACTAAGGTCTGAAACAATCGGAGAAAGAACCCA
          V   S   G   K   R   N   L   W   E   K   Q   S   V   D   K   V   T   S   P   T   K   V   *   531
 2341 ACTGCAAGCCATCTTGCTGGGCCAGCTCAGTTGGAGAGGGCTAATCGCTCTGTTTATATTTATGTTTTCCAATATCCCAGTAAATTCATG
 2431 TATATGCTCACTATATTTAATAACCACAAGTAGAGGTGTTCATGGTCAAAGTGCTGCCTTTGCAGAGGAGCCTGTTTCTAAAGAAACCCA
 2521 TGCTGTGAAGTAAGCTTGCTACTGTCATATGAACAGTGATACCAACCACATCAGAAGTCGACAAAAGAAAATCGAGCTTAAGATTGTCCA
 2611 AGGAATGTATGCGGTATCTAGGTGGGGGGGAAATGACCCATAGGCTTTGTTTTGTCTCAGTACAGACATCATGTTTCTGCACTTTAGACTA
 2701 AAGCATGGAAGAAATTATCTAAGTAGGCAATCAAAATTCTCTGAAAGTGATCCACTTCAGATCTGATATAGGGCAGTGATGATTGTCTTT
 2791 TTTTTTAAAAAAGAAGATGTACTGTTGACATATTGCTTTTCTTCTATGCTGATTCATACCTAGATTGGGTGATTATTTTAGCTGACAGTG
 2881 GTACTGATTTTTTTCTTCAGGTTAGTTGCTTTGTGGATTTCTCTGGTAGCGATAGTAGACTGAACCACATTTAGATATAACCCAACTATG
 2971 TAATGTATGTGCATACGTGTATACAGACACACTAATGGTAGATGACTCTTTCATGCTGGCGCTATTCTATTTCATAGTACGCCCTTGTAA

116

```
3061 CTAACCAATATCACAGCTTTCAAAGATTAAAGAAAATCACAATAGTATATCAATATTTCATATTTGCCAGTAGAAACATGGAAGTTAGGT
3151 ATAGATATGTTTTCAATGCCCAATGACTTATAAGAAAAAAGTATTGGAAAAATAAGAGATTACAAGTGTCAAAACTAGTTGAAACTTCTT
3241 ATAAGACATAGTATTTAGTTTATAATTGAGAGCAGTCTTGAATTCCAGTGTGAATTTTATTAAGTCTACCATCTGGACAAAGCCCAAACC
3331 CTGTTTATTTGTGGCCGTGAAGTTCACCTCCTCCACACACAAAAAAAGTTGACCTACATGATACCTAAGGTGTCTCCTTTCTCTACGAAT
3421 TCCGGAATTCCGGAATTCCCCAACTGCACTGTCTCCACACAGCTTTCCTTCCCACTGGTTACAAGCAAATTGGACAACAAAATCTCATGA
3511 GATATTTGTGATTTAATTTTAGTCACAAAACATCTTCAAAATGACGAGGATTTTGACAGCATGCAAAGTGGTGAAGACACTGAAGAGTGG
3601 CTTTGGTTTGGCCAATGTGACTTCGAAGCGACAGTGGGACTTCTCTAGACCTGGCATCAGGCTCCTTTCTGTGAAGGCACAGACAGCACA
3691 CATTGTTCTGGAAGATGGAACTAAGATGAAGGGCTACTCCTTTGGCCATCCCTCCTCGGTTGCTGGCGAAGTGGTTTTTAATACTGGCTT
3781 AGGAGGGTACTCGGAAGCACTTACTGATCCTGCCTACAAGGGGCAGATCCTCACCATGGCCAACCCTATCATTGGGAATGGTGGGGCCCC
3871 TGATACAACGGCACGTGATGAACTGGGACTGAATAAGTACATGGAGTCTGATGGAATCAAGGTGGCGGGTCTGCTGGTGCTGAATTACAG
3961 TCATGACTACAACCACTGGCTGGCCACCAAGAGTCTGGGCAGTGGCTGCAGGAGGAGAAGGTCCCTGCAATTTATGGAGTGGGTACAAGA
4051 ATGCTGACTAAAATAATTCGGGATAAGGGTACCATGCTTGGGAAGATTGAGTTTGAGGCCAGTCTGTGGACTTTGTGGATCCTAATAAGC
4141 AGAATTTGATTGCCGAGGTTTCAACCAAGGATGTCAAGGTGTTTGGCAAAGGAAACCCCACGAAAGTGGTAGCCGTGGACTGTGGGATAA
4231 AAAACAATGTCATCCGCCTGCTAGTTAAGCGAGGAGCGGAAGTGCATTTGGTCCCCTGGAATCATGACTTCACCCAGATGGACTATGACG
4321 GACTTCTGATCGCTGGAGGACCTGGGAACCCAGCTCTGCACAGCCACTAATTCAGAACGTGAAGAAGATTTTGGAGAGTGACCGCAAAGA
4411 GCCGTTGTTTGGAATCAGTACAGGAAACATAATAACAGGATTGGCTCGTGGCGCCAAATCCTACAAGATGTCCATGGCCAACAGAGGACA
4501 GAACCAACCTGTTTTGAATATCACAAACAGACAGGCTTTCATAACTGCTCAGAATCATGGCTATGCTCTGGACAACACCCTCCCTGCTGG
4591 CTGGAAACCACTGTTTGTCAATGTCAATGACCAAACAAATGAGGTACACGCTCTCAACAATCCATTCAGGTGGAAGTAAGAAGTTATGGC
4681 TTTTCATAGTCTGTTTTATACTTCTTTCCCATGTCTATGTGGTGTGCATGTGCATATGTGCCACACACTCGTGTGTGTGCATGTGCATGT
4771 CTATTGTGTAGGTGTGTGCACCCATGTGTGTCCATGGGGAGCCCGGAGCAGGACACCCGGTTTCTTCCTGGATTGTTCTTCACTTAATTG
4861 TCGTATACATGGCCACTCGTTGAACTGAACCTTTGGCTTGGCTGGCTGGTCAATAAATAAGCTCTCAGGAAGCGTTCAGCTCCATTCCTT
4951 AACCCTGAGGTTATAGCACATGCACTGACCAGCTCCATTCCTTAACGCTGAGGTTATAAACGCATGCACTGACCAGCTCCATTCCTTAAC
5041 GCTGAGGTTATAAACGCATGCACTGACCAGCTCCATTCCTTAACCCTGAGGTTATAGCGCATGCACTGACCACGTCCATTCCTTAACGCT
5131 GAGGTTATAAACGCATGCACTGACCAGCTCCATTCCTTAACCCTGAGGTTATAGCGAATGCACTGACCAGCTCCATTCCTTAACGCTGAG
5221 GTTATAGCGCATGCACTGACCAGCTCCATTCCTTAACCCTGAGGTTATAGCGCATGCACTGACCAGCTCCATTCCTTAACCCTGAGGTTA
5311 TAGCGCATGCACTGACCAGCTCCATTCCTTAACGCTGAGGTTATAAACGCATGCACTGGCATGCTTAGCTTCTCACACAGGTGATAGGAA
5401 CTTGGCCTCAGATCCTTGCCTGTAGGTCAAGTGCTCTAAGCCATGGAACGATCTCCCCAGACAGCCTGTCACTTTAAAAAAGTAATTGCT
5491 CTCTTCCTTTCTAATTAGTCGACTTGTTAAGGGCCAGCACTTGCACCTCCGGAATTCCTGCAGCCCGGGGGATCCACTAGTTCTA
```

Figure 2. DNA sequence of rat nonmuscle caldesmon and its reduced amino acid sequence.

caldesmon is localized diffusely throughout cytoplasm in metaphase. During early stages of cytokinesis, caldesmon is still diffusely present and not concentrated in contractile rings, in contrast to the dramatic accumulation of actin in cleavage furrows during this process. In later stages of cytokinesis, most caldesmon is observed to be yet diffusely localized although some concentration of caldesmon is observed in cortexes as well as in cleavage furrows. When daughter cells begin to spread, caldesmon shows complete co-localization with F-actin-containing structures. Biochemical analyses have shown that changes in the level of microfilament-associated caldesmon during cell division are well correlated with the immunofluorescent observation, and that dephosphorylation of caldesmon appears to precede its re-association with microfilaments. While caldesmon is highly phosphorylated during metaphase, some caldesmon starts to be dephosphorylated at the beginning of cytokinesis. When most cells finish cytokinesis, dephosphorylation is complete. These results suggest that the dissociation of caldesmon during cytokinesis is required in order to allow the specific functions of microfilaments during cell division, including disassembly of stress fibers, and assembly and activation of contractile rings.

One approach toward elucidation of the physiological functions of mitosis-specific phosphorylation of caldesmon would be generating mutant caldesmons lacking the mitosi-specific phosphorylation sites and transfecting such mutants into cells to see what effects on microfilament organization during mitosis these mutants will have. To this end, we have cloned cDNAs encoding rat nonmuscle caldesmon. We have isolated five overlapping cDNA clones from rat liver lambda gt11 cDNA and lambda Zap libraries. These include A18 (1.2kb), A16 (0.8kb), AD24 (2.8kb), D3 (1.6kb), and D43 (3.6kb). These clones were sequenced to generate a 5544-base pair (bp) "contig" including the entire coding region (531 amino acids) as well as 723 bp of 5'-noncoding and 3225 bp of 3'-noncoding regions (Fig. 2). The deduced protein molecular weight is 60,607,which is significantly less than the molecular weight (83,000) estimated by SDS gel electrophoresis. Northern blot analyses, using either A16 (the N-terminus of the coding region) or D3 (the C-terminus of the coding region) as probes, revealed only one size of mRNA (about 5kb) in both smooth muscle and nonmuscle cells.

Rat nonmuscle caldesmon has seven putative phosphorylation sites with consensus sequences (TP or SP) for cdc2 kinase (see Fig. 3 for their positions). Among these, five sites are conserved in many species of caldesmons including rat, chick, and human caldesmons. Oligonucleotide-directed mutagenesis was performed to make mutations at these sites of Ser or Thr to Ala, and such mutants, as well as wild type caldesmons, were expressed in bacteria. After phosphorylation with cdc2 kinase, phosphorylation patterns were analyzed by two-dimensional peptide mapping. In this way we have identified five sites (Ser-462, Thr-468, Ser-491, Ser-497, and Ser-527) for mitosis-specific phosphorylation by cdc2 kinase (manuscript submitted). We are now in the process of transfectiing these mutants using a inducible vector because we found that overexpression of mutant caldesmon as well as wild-type caldesmon is lethal. The preliminary results have shown that overexpression of caldesmon initially caused the stabilization of stress fibers. Further expression, however, caused disorganization of microfilaments including disassembly of stress fibers, resulting in cell rounding and death.

Fig. 3 shows the domain structure of rat nonmuscle caldesmon with the locations of phosphorylation sites, as well as actin- calmodulin- and myosin-binding domains indicated. The assignments of myosin-, actin-, and calmodulin-binding domains of rat nonmuscle caldesmon are adopted from the domain structure of chick smooth muscle caldesmon determined by others (see, for review, ref[7]). It is worthy of note that all phosphorylation sites (Ser-462, Thr-468, Ser-491, and Ser-497) except one (Ser-527) are localized outside the actin and calmodulin binding domains. This is consistent with our finding that all mutants lacking phosphorylation sites show normal actin and calmodulin binding, and suggests that these phsophorylation sites interact with or otherwise modify accessibility of the actin and calmodulin binding domains.

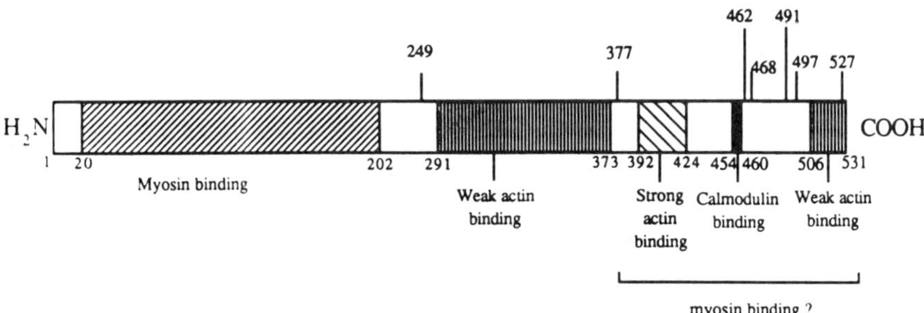

Figure 3.   Domain structure of rat nonmuscle caldesmon. The domain structure suggested for chick smooth muscle caldesmon by others (see, for example, ref[7,37]) are adopted to assign myosin-, actin-, and calmodulin-binding domains of rat nonmuscle caldesmon. Numbers on the top of the figure show phosphorylation sites containing either SP or TP. Of these, Ser-462, Thr-468, Ser-491, Ser-497, and Ser-527 are identified as the phosphorylation sites by cdc2 kinase.

CALDESMON IN CELL TRANSFORMATION

Caldesmon appears to control the stability of microfilament structure through the regulation of the actin binding affinity of tropomyosin. Tropomyosin stabilizes the structure of microfilaments. We, as well as others, demonstrated that the relative levels of expression of multiple tropomyosin isoforms change concomitantly with the morphological alterations associated with cell transformation (for review, see ref[38]). Low Mr tropomyosin isoforms prominent in transformed cells show weaker affinity for actin than the predominantly high Mr tropomyosin isoforms found in normal cells[39] This phenomenon appears to be directly related to the instability of microfilament structures in transformed cells.

We have found that caldesmon stimulates actin binding by tropomyosin by as much as 10-fold[12], and that caldesmon together with tropomyosin inhibits actin severing and capping activities of gelsolin. Furthermore, caldesmon has been reported to show reduced expression in transformed cells[5,6]. These changes in the expresion of tropomyosin and caldesmon thus imply that microfilaments are destabilized in transformed cells.

Previous work has demonstrated that the mRNA level of caldesmon is reduced in several types of human transformed cells[5]. However, such study was performed with cell lines with different genetic backgrounds. We have thus examined the levels of caldesmon mRNA using rat-1 cells transformed with a RSV mutant, temperature-sensitive for transformation. These cells are isogenic. They have a well spread morphology at 39°C, similar to that of normal rat-1 cells, but at 34°C they display a transformed phenotype, namely a rounded morphology. The levels of caldesmon mRNA were examined by Northern blotting using the cDNA insert from A16 as a probe, which hybridized to a single 5-kb band of Rat-1 mRNA. As Fig. 4 shows, caldesmon mRNA levels showed at least a 2-fold decrease upon cell transformation resulting from the temperature shift from 39°C to 34°C. Levels of actin mRNA, measured as a control, did not decrease upon temperature shift-down. These results suggest that the decrease in a caldesmon levels observed upon cell transformation is caused by reduced synthesis of caldesmon mRNA and/or increased mRNA turnover.

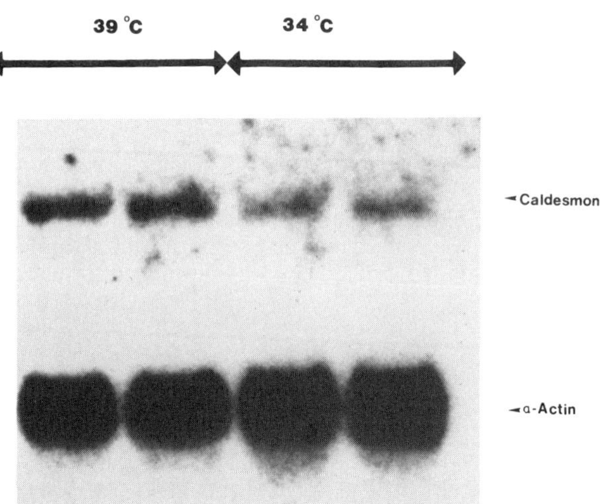

Figure 4   Northern blotting analyses of mRNA levels of caldesmon in rat-1 cells transformed with a Rous sarcoma virus mutant, temperature-sensitive for transformation.

POSSIBLE FUNCTIONS OF CALDESMON

Our studies have demonstrated that alterations in microfilament patterns associated with either cell transformation or cell division lead to the same biochemical changes, i. e. inhibition of caldesmon's binding to microfilaments. This reduction of caldesmon binding is achieved in two different ways. Upon cell transformation, the expression of caldesmon is decreased, thereby decreasing the amount of caldesmon bound to actin. During mitosis, caldesmon is disassociated from microfilaments as a result of mitosis-specific phosphorylation, while the total cellular levels of caldesmon are unchanged. These observations strongly suggest a key role for caldesmon in mediating the alterations of microfilament patterns during these two biological activities.

The inhibition of caldesmon binding may affect the contractility of the actomyosin system during mitosis and in cell transformation. Previous work has shown that caldesmon inhibits actin-activated myosin ATPase activity[22,24-26,40,41]. Ca^{2+}/calmodulin releases caldesmon's inhibition of myosin ATPase as it reduces the actin binding affinity of caldesmon. Our *in vitro* studies have shown that the dissociation of caldesmon from microfilaments resulting from its mitosis-specific phosphorylation also releases its inhibition of actomyosin ATPase[42]. This release may lead to contraction of the actin-myosin system in mitotic cells in the following two ways: First, such contraction may cause rounding-up of cell shape when cells enter prophase. Second, the dissociation of nonmuscle caldesmon may be needed at a later stage to activate contractile rings during cytokinesis (Indeed, we have found that caldesmon is absent from contractile rings during cytokinesis). In cell transformation, on the other hand, the total level of caldesmon is decreased independent of the cell cycle. This reduction may cause unregulated contraction of the actomyosin system in transformed cells, resulting in maintenance of rounded morphologies.

Caldesmon may also function in the reorganization of microfilaments by regulating gelsolin activity. As described previously, we have shown that caldesmon coupled with tropomyosin inhibits both severing and capping activities of gelsolin[8,9]. The dissociation of caldesmon from microfilaments by mitosis-specific kinase activity may thus release the caldesmon mediated inhibition of gelsolin activity, resulting in the severing of microfilaments. This severing may lead to the disassembly of stress fibers and the concomitant morphological alterations seen during prophase. Similar but unregulated activation of gelsolin may occur in transformed cells due to the reduction of caldesmon levels upon transformation. In addition, tropomyosin isoforms with lower actin binding affinities become predominant in transformed cells[43-49], resulting in further activation of gelsolin's actin severing and capping activities. These changes may cause permanent morphological changes and produce the rounded cell shapes seen in many transformed cells. It seems plausible that changes in the expression of caldesmon and tropomyosin genes are involved in permanent morphological alterations, such as those associated with development or cell transformation, while post-translational modification (phosphorylation) may control transient morphological alterations such as those occurring during mitosis. The observation that both mitotic cells and transformed cells not only share similar rounded morphologies but also show similar biochemical changes (including the reduction of actin binding by caldesmon) suggests that the reorganization of microfilaments that causes rounded cell shapes may be related to the rapid cell cycling of cancer cells.

REFERENCES

1. R Pollack, M Osborn, and K Weber, Patterns of organization of actin and myosin in normal and transformed cultured cells, *Proc. Natl. Acad. Sci. U. S. A.* 72:994(1975).
2. JW Sanger, JM Sanger Actin localization during cell division. in:Cell Motility R Goldman, T Pollard, J Rosenbaum, eds., Cold Spring Harbor Laboratory, Cold Spring Harbor, NY, 1295.(1976)
3. TE Schroeder Actin in dividing cells: evidence for its role in cleavage but not mitosis. in:Cell Motility R Goldman, T Pollard, J Rosenbaum, eds., Cold Spring Harbor Laboratory, Cold Spring Harbor, NY, 265.(1976)
4. S Yamashiro, Y Yamakita, R Ishikawa, and F Matsumura, Mitosis-specific phosphorylation causes 83K non-muscle caldesmon to dissociate from microfilaments, *Nature* 344:675(1990).
5. RE Novy, JL Lin, and JJ Lin, Characterization of cDNA clones encoding a human fibroblast caldesmon isoform and analysis of caldesmon expression in normal and transformed cells, *J. Biol. Chem.* 266:16917(1991).
6. M Koji-Owada, A Hakura, K Iida, I Yahara, K Sobue, and S Kakiuchi, Occurrence of caldesmon (a calmodulin-binding protein) in cultured cells: Comparison of normal and transformed cells, *Proc. Natl. Acad. Sci. USA* 81:3133(1984).
7. K Sobue and JR Sellers, Caldesmon, a novel regulatory protein in smooth muscle and nonmuscle actomyosin systems, *J. Biol. Chem.* 266:12115(1991).

8. R Ishikawa, S Yamashiro, and F Matsumura, Differential modulation of actin-severing activity of gelsolin by multiple isoforms of cultured rat cell tropomyosin. Potentiation of protective ability of tropomyosins by 83-kDa nonmuscle caldesmon, *J. Biol. Chem.* 264:7490(1989).

9. R Ishikawa, S Yamashiro, and F Matsumura, Annealing of gelsolin-severed actin fragments by tropomyosin in the presence of Ca2+. Potentiation of the annealing process by caldesmon, *J. Biol. Chem.* 264:16764(1989).

10. K Sobue, Y Muramoto, M Fujita, and S Kakiuchi, Purification of a calmodulin-binding protein from chicken gizzard that interacts with F-actin, *Proc. Natl. Acad. Sci. USA* 78:5652(1981).

11. A Bretscher and W Lynch, Identification and localization of immunoreactive forms of caldesmon in smooth and nonmuscle cells: a comparison with the distributions of tropomyosin and alpha-actinin, *J. Cell Biol.* 100:1656(1985).

12. PE Hoar, WGL Kerrick, and PS Cassidy, Chicken Gizzard:Relation between calcium-activated phosphorylation and contraction, *Science* 204:503(1979).

13. F Matsumura and S Yamashiro-Matsumura, Purification and characterization of multiple isoforms of tropomyosin from rat cultured cells, *J. Biol. Chem.* 260:13851(1985).

14. K Sobue, T Tanaka, K Kanda, N Ashino, and S Kakiushi, Purification and characterization of caldesmon77: a calmodulin-binding protein that interacts with actin frilaments from bovine adrenal medulla, *Proc. Natl. Acad. Sci. USA* 82:5025(1985).

15. S Yamashiro-Matsumura, R Ishikawa, and F Matsumura, Purification and characterization of 83 kDa nonmuscle caldesmon from cultured rat cells: Changes in its expression upon L6 myogenesis, *Protoplasma* Suppl. 2:9(1988).

16. Y Yamakita, S Yamashiro, and F Matsumura, Microinjection of nonmuscle and smooth muscle caldesmon into fibroblasts and muscle cells, *J. Cell Biol.* 111:2487(1990).

17. J Bryan, M Imai, R Lee, P Moore, RG Cook, and WG Lin, Cloning and expression of a smooth muscle caldesmon, *J. Biol. Chem.* 264:13873(1989).

18. K Hayashi, Y Fujio, I Kato, and K Sobue, Structural and functional relationships between h- and l-caldesmons, *J. Biol. Chem.* 266:355(1991).

19. T Fujii, M Imai, GC Rosenfeld, and J Bryan, Domain mapping of chickcn gizzard caldesmon, *J. Biol. Chem.* 262:2757(1987).

20. VM Riseman, WP Lynch, B Nefsky, and A Bretscher, The calmodulin and F-actin binding sites of smooth muscle caldesmon lie in the carboxyl-terminal domain whereas the molecular weight heterogeneity lies in the middle of the molecule, *J. Biol. Chem.* 264:2869(1989).

21. M Urban:c:ikov:a and M Gröfov:a, Quantitative changes of tropomyosin synthesis in RSV-transformed mammalian cells in comparison with normal fibroblasts, *Neoplasma.* 29:655(1982).

22. CW Smith, K Pritchard, and SB Marston, The mechanism of Ca2+ regulation of vascular smooth muscle thin filaments by caldesmon and calmodulin, *J. Biol. Chem.* 262:116(1987).

23. DB Sacks and JM McDonald, Effects of cationic polypeptides on the activity, substrate interaction, and autophosphorylation of casein kinase II: a study with calmodulin, *Arch. Biochem. Biophys.* 299:275(1992).

24. KY Horiuchi, H Miyata, and S Chacko, Modulation of smooth muscle actomyosin ATPase by thin filament associated proteins, *Biochem. Biophys. Res. Commun.* 136:962(1986).

25. R Dabrowska, A Goch, B Galazkiewiecz, and H Osinska, The infuluence of caldesmon on ATPase activity of the skeletal muscle actomyosin and bundling of actin filaments, *Biochim. Biophys. Acta* 842:70(1985).

26. ME Hemric and JM Chalovich, Effect of caldesmon on the ATPase activity and the binding of smooth and skeletal myosin subfragments to actin, *J. Biol. Chem.* 263:1878(1988).

27. B Galazkiewicz, J Belagyi, and R Dabrowska, The effect of caldesmon on assembly and dynamic properties of actin, *Eur. J. Biochem.* 181:607(1989).

28. B Galazkiewicz, F Buss, BM Jockusch, and R Dabrowska, Caldesmon-induced polymerization of actin from profilactin, *Eur. J. Biochem.* 195:543(1991).

29. I Mabuchi and M Okuno, The effect of myosin antibody on the division of starfish blastomeres, *J. Cell Biol.* 74:251(1977).
30. D Knecht and WF Loomis, Antisense RNA inactivation of myosin heavy chain gene expression in *Dictyostelium discoideum, Science* 236:1081(1987).
31. A DeLozanne and JA Spudich, Disruption of the *Dictyostelium* myosin heavy chain gene by homologous recombination, *Science* 236:1086(1987).
32. LA Shuvalova, MV Ostrovskaia, EA Sosunov, and VV Lednev, [The effect of a weak magnetic field in the paramagnetic resonance mode on the rate of the calmodulin-dependent phosphorylation of myosin in solution], *Dokl. Akad. Nauk. SSSR.* 317:227(1991).
33. M Peter, J Nakagawa, M Doree, JC Labbe, and EA Nigg, In vitro disassemblyu of the nuclear lamina and M-phase specific phosphorylation of lamins by cdc2 kinase, *Cell* 61:591(1990).
34. Y-H Chou, JR Bischott, D Beach, and RD Goldman, Intermediate filament reorganization during mitosis is mediated by p34^{cdc2}; Phosphorylation of vimentin, *Cell* 62:1063(1990).
35. F Matsumura and S Yamashiro-Matsumura, Modulation of actin-bundling activity of 55-kDa protein by multiple isoforms of tropomyosin, *J. Biol. Chem.* 261:4655(1986).
36. N Hosoya, H Hosoya, S Yamashiro, H Mohri, and F Matsumura, Localization of caldesmon and its dephosphorylation during cell division, *J. Cell Biol.* 121:1075(1993).
37. CL Wang, LW Wang, SA Xu, RC Lu, V Saavedra-Alanis, and J Bryan, Localization of the calmodulin- and the actin-binding sites of caldesmon, *J. Biol. Chem.* 266:9166(1991).
38. F Matsumura and S Yamashiro-Matsumura, Tropomyosin in cell transformation, *Cancer Rev.* 6:21(1986).
39. F Matsumura and S Yamashiro-Matsumura, Purification and characterization of multiple isoforms of tropomyosin from rat cultured cells, *J. Biol. Chem.* 260:13851(1985).
40. K Sobue, K Takahashi, and I Wakabayashi, Caldesmon150 regulates the tropomyosin-enhanced actin-myosin interaction in gizzard smooth muscle, *Biochem. Biophys. Res. Commun.* 132:645(1985).
41. K Pritchard and SB Marston, Ca2+-calmodulin binding to caldesmon and the caldesmon-actin-tropomyosin complex. Its role in Ca2+ regulation of the activity of synthetic smooth-muscle thin filaments, *Biochem. J.* 257:839(1989).
42. Y Yamakita, S Yamashiro, and F Matsumura, Characterization of mitotically phosphorylated caldesmon, *J. Biol. Chem.* 267:12022(1992).
43. F Matsumura, JJ-C Lin, S Yamashiro-Matsumura, GP Thomas, and WC Topp, Differential expression of tropomyosin forms in the microfilament isolated from normal and transformed rat cultured cells, *J. Biol. Chem.* 258:13954(1983).
44. F Matsumura, S Yamashiro-Matsumura, and JJ-C Lin, Isolation and characterization of tropomyosin-containing microfilaments from cultured cells, *J. Biol. Chem.* 258:6636(1983).
45. JJ-C Lin, S Yamashiro-Matsumura, F Matsumura Microfilaments in normal and transformed cells: Changes in the multiple forms of tropomyosin. in:Cancer Cells. 1 A Levine, G Vande Woude, WC Topp, J Watson, eds., Cold Spring Harbor Laboratory, New York, 57.(1984)
46. M Hendricks and H Weintraub, Multiple tropomyosin polypeptides in chicken embryo fibroblasts: differential repression of transcription by Rous sarcoma virus transformation, *Mol. Cell Biol.* 4:1823(1984).
47. M Hendricks and H Weintraub, Tropomyosin is decreased in transformed cells, *Proc. Natl. Acad. Sci. U. S. A.* 78:5633(1981).
48. CL Leonardi, RH Warren, and RW Rubin, Lack of tropomyosin correlates with the absence of stress fibers in transformed rat kidney cells, *Biochim. Biophys. Acta* 720:154(1982).
49. JJ Lin, DM Helfman, SH Hughes, and CS Chou, Tropomyosin isoforms in chicken embryo fibroblasts: purification, characterization, and changes in Rous sarcoma virus-transformed cells, *J. Cell Biol.* 100:692(1985).

CYTOSKELETON, MOTILE STRUCTURES AND MACROMOLECULAR CROWDING

Enrico Grazi

Istituto di Chimica Biologica,
Università di Ferrara,
Ferrara, Italy

INTRODUCTION

We descrive the effect of macromolecular crowding on the associations of the cytoskeletal and of the motile structures. These effects are due to preferential interactions and have been treated theoretically by many authors[1,2,3,4]. Occasionally, preferential interactions are taken into account in the study of the biological reactions, their role, however, is not yet adequately recognized.

Macromolecular crowding shifts the monomer-tetramer equilibrium of glyceraldehyde-3-phoshate dehydrogenase in favour of tetramer formation[5]; allows blunt-end ligation by DNA ligase from rabbit liver and Escherichia coli[6]; drives catenation of supercoiled and gapped DNA circles by topoisomerase I[7]; regulates the transporter-mediated ion flux across cell membranes[8]. In the specific field of the cytoskeleton preferential interactions influence the binding of glycolytic enzymes to cytoskeletal structures[9] and myofibrils[10]; accelerate the rate and increase the extent of actin polymerization[11]; increase the rate of elongation of actin filaments[12]; favour the formation of bundles of actin filaments[13].

This last contribution stimulated our interest in the topic. We began a systematic investigation and discovered that cytoskeletal functions are widely influenced, both quantitatively and qualitatively, by large concentrations of macromolecular solutes. The influence is spanning from the reversible interconversion of actin filaments into actin bundles, to the intensification of the actin gelling activity of alpha-actinin, to the lowering of the solating activity of gelsolin, to the triggering of actomyosin retraction, supported by unphosphorylated smooth muscle myosin. All these effects are briefly presented and discussed.

EXPERIMENTAL

The effect of macromolecular crowding on the formation and the dissociation of actin bundles, as well as on the solating activity of gelsolin was studied at pH 7.0 and 37°C. Final electrolyte concentration was as follows : 107 mM K$^+$, 13 mM Na$^+$, 6 mM Mg^{2+}, 5 mM orthophosphate, 4 mM ATP, 37 mM Cl$^-$ and 70 mM propionate, ionic strength was O.145 M[18,19,29,47].

Actin: Biophysics, Biochemistry, and Cell Biology
Edited by J.E. Estes and P.J. Higgins, Plenum Press, New York, 1994

The actin gelling activity was studied at 37°C and pH 7.5. Final electrolyte concentration was : 0.1 M KCl, 2 mM $MgCl_2$, 0.5 mM ATP, 10 mM Tris-HCl[42].

Unphosphorylated smooth muscle myosin was studied at 37°C and pH 7.0. Final electrolyte concentration was : 148 mM K^+, 13 mM Na^+, 15.5 mM Mg^{2+}, 9.5 mM sulfate, 6 mM phoshate, 15 mM creatine phosphate, 6.5 mM ATP, 66 mM propionate[51].

Macromolecular crowding was mimicked either by poly(ethylene glycol) 6000 or by serum albumin.

CELL VOLUME AND MACROMOLECULAR CROWDING

Isosmotic changes of cell volume are induced by tumor promoters and by mitotic agents, they are operated by K-Cl cotransport or by Na-H exchange, are accompanied by osmotically obliged water and, consequently, by the change of the concentration of the macromolecules[14]. Being produced by water fluxes, these changes propagate quite rapidly to all the cellular space, included the nuclear matrix, where they may act as signals for gene transcription.

Small fractional changes of cell volume, by generating small changes of the concentration of the macromolecules of the medium, may modify the state of aggregation of those actin populations, that are at the transition between filaments and bundles.

Conversely, changes in the state of aggregation of actin may influence the volume. In neutrophils, F-actin content undergoes cyclic variations, that correlate with periodic fluctuations in lamellipod size : actin polymerization (and possibly bundles formation) corresponds closely to lamellipod extension; actin depolymerization corresponds closely to lamellipod retraction[15]. The polymerization of actin is assumed to drive the extension of lamellipods as well as the extension of the acrosomal process of Thyone sperm[15,16].

Many animal cell types respond to swelling or shrinkage by activating membrane transporters : the swelling-induced K-Cl cotransport, the shrinkage-induced Na-H exchange. The small fractional changes in cell volume are sensed as small fractional changes in the cytoplasmic macromolecular concentration. These, in turn, elicit an amplified change of the thermodynamic activity of proteins, that regulate membrane transport[8]. The scaled particle theory[17] predicts, in fact, that, in a highly crowded solution, the activity coefficient of each macromolecular species is a function of the total concentration of the macromolecules.

ANCILLARY CYTOSKELETAL PROTEINS AND THE CONVERSION OF ACTIN FILAMENTS INTO

ACTINS BUNDLES

The ancillary cytoskeletal proteins modulate the concentration of the macromolecules required to promote the transition of actin filaments into actin bundles. As an example, in our experimental conditions, the formation of bundles of actin filaments occurs at 3% poly(ethylene glycol) for caldesmon-decorated actin, at 4-5% poly(ethylene glycol) for filamin-decorated actin, at 5-7% poly(ethylene glycol) for caldesmon-tropomyosin-decorated actin, at 6-7% poly(ethylene glycol) for F-actin and at 9-10% poly(ethylene glycol) for tropomyosin-decorated actin[18,19].

A finer regulation is achieved by changing the ratios of the ancillary proteins with respect to actin. Filamin-decorated actin undergoes bundling, even in the absence of poly(ethylene glycol), when the filamin to actin molar ratio is increased to 1:8[20]. The concentration of poly(ethylene glycol), required to induce the transition of F-actin filaments into actin bundles, decreases with the increase of the caldesmon to actin molar ratio, but, even at a caldesmon to actin molar ratio of 1:3, poly(ethylene glycol) is required for bundling[19].

The intracellular distribution of ancillary cytoskeletal proteins may determine the state of aggregation of actin. Filamin, that favours actin bundling, is present in the leading edge and membrane ruffles[20,21], while tropomyosin, that hinders actin bundling, is confined to the internal part of the cultured cells[22].

In some cases the interplay between macromolecular concentration, state of aggregation of actin and distribution of the ancillary cytoskeletal proteins is quite complex. Caldesmon is reported to bind tighter to tropomyosin-decorated F-actin than to F-actin[23], as a consequence, a system composed by tropomyosin-decorated F-actin and by caldesmon-decorated F-actin evolves toward a system composed by F-actin and by tropomyosin-caldesmon-decorated F-actin. This has a profound effect on the state of aggregation of actin. In 3% poly(ethylene glycol) solutions, the progression of the system from the first to the second state increases the proportion of actin filaments over actin bundles. In more concentrated poly(ethylene glycol) solutions (6-7%), the same progression of the system increases the proportion of actin bundles over actin filaments[19].

MECHANISMS FOR THE DISSOCIATION OF ACTIN BUNDLES

Many scientists, being used to work in solutions of small electrolytes, may consider the filamentous form as the ground state of aggregation of actin in the cell. In this respect, the report of Suzuki et al.[13] is quite striking. These authors show that actin bundles represent the ground state of aggregation of actin, when the osmolarity of the macromolecules of the medium is close to that of the cell sap. The question to ask , therefore, is how, in the cell, actin bundles can be forced to dissociate into actin filaments.

Many actin bundling proteins are proposed to regulate the rapid formation and dissociation of actin bundles.

Human erythrocyte protein 4.9 is reported to display an actin-bundling activity, which is abolished when the protein, in preformed actin bundles, is phosphorylated by the catalytic subunit of cyclic AMP-dependent protein kinase[24].

Lipocortin 85 is reported to display a Ca^{2+}-dependent actin-bundling activity[25].

The actin-bundling activity of synapsin I is reported to be abolished, when the protein is phosphorylated by ca^{2+}-calmodulin-dependent protein kinase II[26].

The Ca^{2+}-calmodulin-caldesmon complex is reported to dissociate at low Ca^{2+} concentration; in turn, the released caldesmon is reported to displace filamin from filamin-actin bundles, which then dissociate into filaments[27,28].

In no cases, however, it was checked whether preformed actin bundles actually dissociate, when the concentration of the macromolecules is close to that present in the cytosol. As an example, we found that, in 3% poly(ethylene glycol), Ca^{2+}-calmodulin prevents the binding of caldesmon to F-actin but fails to dissociate caldesmon-F-actin bundles, that are formed at that concentration of poly(ethylene glycol) (unpublished results). This is at variance with the data of Sobue et al.[27], obtained in solutions of small electrolytes.

When the dissociation of actin bundles is operated by a protein, the main drawback is the steric hindrance, due to the dense packing of actin bundles. As an example, in resting solutions, tropomyosin does not dissociate F-actin bundles in 7% poly(ethylene glycol), eventhough tropomyosin-decorated F-actin forms bundles only at 9-10% poly(ethylene glycol)[18,19]. To achieve the dissociation of the bundles the solution must be constantly mixed[29].

On the contrary, even in resting solutions, tropomyosin promotes the dissociation of caldesmon-decorated F-actin bundles in 3% poly(ethylene glycol). Probably, in this case, actin bundles are less densely packed[19], because of the lower osmotic stress, caused by the lower concentration of poly(ethylene glycol).

It is unlikely that cytoplasmic streaming may overcome the steric constraints, imposed by the densely packed bundles of actin filaments : it thus appears that dissociation of actin bundles is best triggered by small molecules. A potential candidate to this role is Mg^{2+}. In vitro, the increase of free Mg^{2+} concentration favours bundling of actin, while the decrease of free Mg^{2+} concentration favours the dissociation, into filaments, of the tropomyosin-decorated F-actin system. F-actin alone is insensitive to this mechanism of regulation[29].

At 37°C, pH 7.14, in 7,2% poly(ethylene glycol), the transition between filaments and bundles occurs at 1.7 - 2.0 mM free Mg^{2+}. These concentrations are quite larger than 0.6 mM, that is the estimated value for free Mg^{2+} concentration in mammalian tissues[30]. Nevertheless, because of the multiplicity of the factors that regulate the transition between actin filaments and actin bundles and because of the practical impossibility to match cell conditions, the free Mg^{2+} concentration cannot be excluded as a potential candidate for regulation. Moreover, it must be taken into account that, in the cell, the concentration of free Mg^{2+} may be increased by lowered energy charge of the ATP system[31], by altered intracellular Mg^{2+} binding, secondary to changes in Ca^{2+} influx, or pH or to cell shrinkage[32]. In our experimental conditions a shift from 1.7 to 2 mM free Mg^{2+}, which produces a substantial increase in the amount of tropomyosin-decorated actin bundles over tropomyosin-decorated actin filaments, is brought about by a 0.5 mM decrease in ATP concentration. Stimulation by ADP of human platelets induces a similar decrease in "metabolic" ATP concentration[32]. These changes in ATP concentration are accompanied by the formation of actin bundles in the course of the aggregation reaction[33,34].

THE ACTIN GELLING ACTIVITY OF ALPHA-ACTININS AND MACROMOLECULAR CROWDING

It is a common notion that the actin gelling activity of alpha-actinins becomes almost undetectable at 37°C[35,36,37,38,39]. This behaviour, apparently, casts doubt on the functioning of these proteins in vivo[37]. As a matter of fact, alpha-actinin from chicken gizzard, at nM concentrations, increases significantly the rigidity of actin gel, even at 37°C, provided that the concentration of actin is low (2-3 μM)[40,41]. This finding, that may be regarded as a curiosity, indicates that, only at low concentrations, actin filaments are free to diffuse and to be optimally crosslinked by alpha-actinin. More significantly, at 37°C, in 6% poly(ethylene glycol), addition of 30 nM alpha-actinin to 12 μM actin, induces the gelation of the system and increases the rigidity from 23.5 to 54 dynes/cm^2. The onset of gelation is concomitant with the transition of actin filaments into actin bundles[42]. The binding isotherm of alpha-actinin changes from the anomalous binding isotherm with filamentous actin[40,41] to the hyperbolic binding isotherm ($K_{diss} =$ 11.3 μM) with actin bundles[43]. It is likely that the parallel arrays of filaments, in actin bundles, offer an ordered matrix of sites, which favour the bidentate binding of alpha-actinin. The crosslinking by alpha-actinin prevents the filaments from sliding in actin bundles. As a result, since the network of actin bundles is largely anastomosed, the rigidity of the system is increased by alpha-actinin, even at 37°C. These observations support the view that, in vivo, alha-actinin functions are mostly carried on by interaction with actin bundles, in agreement with the finding that, in the cell, alpha-actinin is mostly associated with actin bundles[44].

GELSOLIN AND MACROMOLECULAR CROWDING

The gelsolin:actin molar ratio approaches 1:100 in the cell[45], a value adequate, in vitro, to support nucleation, capping and cutting of actin filaments. Moreover, since these functions are regulated by Ca^{2+} and by phosphatidylinositol 4,5-bisphosphate[46], gelsolin becomes an attractive candidate to promote gel-sol transition in vivo.

The functions of gelsolin, in vitro, are usually tested on filamentous actin. It is known, however, that, owing to the large concentration of the macromolecules, in the cell actin is often present as bundles of filaments. In these structures, the strong latero-lateral interactions between actin filaments could counteract the effect of filament cutting by gelsolin. This occurs indeed[47]. In the presence of 15 nM gelsolin, rigidity of 12 μM F-actin drops from 3.9 to 0.27 dynes/cm^2, while the rigidity of actin bundles, formed in 6% poly(ethylene glycol), is unaffected, being 29 and 27.5 dynes/cm^2, respectvely, in the absence and in the presence of 30 nM gelsolin. Protection against the action of gelsolin is complete also when the system, in poly(ethylene glycol), is supplemented with 1.5 μM tropomyosin, a condition in which actin bundles are not formed. Perhaps, also in this case, the extent of the latero-lateral association of the filaments increases, eventhough not to such a level to be unambiguosly detected at the electron microscope. The action of gelsolin could also be hampered by the strengthening of monomer-monomer interactions, that occurs in the presence of high concentrations of macromolecules[11].

UNPHOSPHORYLATED SMOOTH MUSCLE MYOSIN AND MACROMOLECULAR CROWDING

It is known that, in vitro, MgATP dissociates unphosphorylated smooth muscle myosin filaments, by promoting the transition of monomeric myosin from the extended (6S) to the folded (11S) shape[48]. In vivo, however, unphosphorylated smooth muscle myosin filaments are stable[49,50], as are stable in vitro, provided that the solutions are supplemented with large concentrations of macromolecules : 6% poly(ethylene glycol) or 25% serum albumin. Thus macromolecular crowding, by driving the association of myosin monomers, displaces the 6S - 11S equilibrium[51].

Ca^{2+}-dependent phosphorylation of the 20 kDa light chain subunit of myosin clearly represents a major pathway for the regulation of contractile force in smooth muscle. As an example, light chain phosphorylation has been shown to initiate both short and sustained arterial contraction[52]. However, many examples of dissociation between force levels, myosin light chain phosporylation levels, intracellular free Ca^{2+} concentration levels and cross-bridge cycling rates have been reported, and the possible existence of a second regulatory mechanism for smooth muscle tone is difficult to negate.   The observation that, in vitro, in the presence of a large concentration of macromolecules, unphosphorylated smooth muscle myosin supports actomyosin retraction, shows that the basic machinery of smooth muscle does not require myosin light chain phosphorylation to initiate contraction[51]. It is very likely that myosin light chain phosphorylation is required only by the fully regulated contractile machinery [51].

Caldesmon may play a relevant role in regulation . The addition of supraphysiological concentrations of caldesmon to skinned gizzard smooth muscle fibers, induces relaxation of submaximal contraction, in the absence of changes in myosin light chain phosphorylation[53]. Conversely, a caldesmon proteolytic fragment, that contains the residues from Gly[651] to Ser[667] plus an added cysteine at the C terminus and that does not inhibit actomyosin ATPase activity, induces contraction in vascular smooth muscle cells of ferret aorta, even at a pCa = 9. The contraction is induced by displacing the inhibtory region of endogenous caldesmon. These results suggest that caldesmon regulates contraction by providing a basal resting inhibition of muscular tone[54].

Acknowledgment : This work was supported by the grants 91/40/05/016 and   91/60/05/033 of the Italian Ministero della Ricerca Scientifica.

REFERENCES

1. A.G. Ogston, Some thermodynamic relationships in ternary systems, with special reference to the properties of systems containing hyaluronic acid and protein, *Archiv. Biochem. Biophys. Suppl.* 1:39(1962).

2. T. Arakawa and S.N. Timasheff, Mechanism of poly(ethylene glycol) interaction with proteins, *Biochemistry* 24:6756(1985).

3. A.P. Minton, Excluded volume as a determinant of macromolecular structure and reactivity, *Biopolymers* 20:2093(1981).

4. J.C. Lee, K. Gekke, and S. Timasheff, Measurements of preferential solvent interactions by densimetric techniques, *Methods Enzymol.* 61:26(1979).

5. A.P. Minton and J. Wilf, Effect of macromolecular crowding upon the structure and function of an enzyme : glyceraldehyde 3-phosphate dehydrogenase, *Biochemistry* 20:4821(1981).

6. S.B. Zimmermann and B.H. Pfeiffer, Macromolecular crowding allows blunt- end ligation by DNA ligases from rat liver or Escherichia coli, *Proc. Natl. Acad. Sci. USA* 80:5852(1983).

7. R.L. Low, J.M. Kaguni, and A. Kornberg, Potent catenation of supercoiled and gapped DNA circles by topoisomerase I in the presence of a hydrophilic polymer, *J. Biol. Chem.* 259:4576(1984).

8. A.P. Minton, G. Craig Colclasure, and J.C. Parker, Model for the role of macromolecular crowding in regulation of cellular volume, *Proc. Natl. Acad. Sci. USA* 89:10504(1992).

9. K. Shearwin, C. Nanhua, and C. Masters, The influence of molecular crowding on the binding of glycolytic enzymes to cytoskeletal structures, *Biochem. Int.* 19:723(1989).

10. S.J. Harris and D.J. Winzor, Effect of thermodynamic nonideality on the subcellular distribution of enzymes : adsorption of aldolase to muscle myofibrils, *Archiv. Biochem. Biophys.* 243:598(1985).

11. R.L. Tellam, M.J. Sculley, L.W. Nichol, and P.R. Wills, The influence of poly(ethylene glycol) 6000 on the properties of skeletal muscle actin, *Biochem.J.* 213:651(1983).

12. D. Drenckhahn and T.D. Pollard, Elongation of actin filaments is a diffusion-limited reaction at the barbed end and is accelerated by inert macromolecules, *J. Biol. Chem.* 261:12754(1986).

13. A. Suzuki, M. Yamazaki, and T. Ito, Osmoelastic coupling in biological structures : formation of parallel bundles of actin filaments in a crystalline-like structure caused by osmotic stress, *Biochemistry* 28:6513(1989).

14. S. Grinstein and J.K. Foskett, Ionic mechanism of cell volume regulation in leukocytes, *Annu. Rev. Physiol.* 52:399(1990).

15. M.P. Wymann, P. Kernen, T. Bengtsson, T. Andersson, M. Baggiolini, and D . A . Derenleau, Corresponding oscillations in neutrophil shape and filamentous actin content, *J. Biol. Chem.* 265:619(1990).

16. L.G. Tilney and S. Inoue, Acrosomal reaction of thyone sperm. II. the kinetics and possible mechanism of acrosomal process elongation, *J. Cell Biol.* 93:820(1982).

17. J.L. Lebovitz, E. Helfand, and E. Prestgaard, Scaled particle theory of fluid mixtures, *J. Chem. Phys.* 43:774(1965).

18. E. Grazi, G. Trombetta, and M. Guidoboni, Divergent effects of filamin and tropomyosin on actin filaments bundling, *Biochem. Biophys. Res. Communs.* 167:1109(1990).

19. P. Cuneo, E. Magri, A. Verzola, and E. Grazi, Macromolecular crowding is a primary factor in the organization of the cytoskeleton, *Biochem. J.* 281:507(1992).

20. K. Wang and S.J. Singer, Interaction of filamin with F-actin in solution, *Proc. Natl. Acad. Sci. USA* 74:2021(1977).

21. J.W. Small, G. Rinnerthaler, and H. Hinnsen, Organization of actin meshworks in cultured cells : the leading edge, *Cold Spring Harbour Symp. Quant. Biol.* 46:599(1982).

22. E. Lazarides, Two general classes of cytoplasmic actin filaments in tissue culture cells: the role of tropomyosin, *J. Supramol. Struct.* 5:531(1976).

23. C.J. Moody, S.B. Marston, and C.W.J. Smith, Bundling of aorta caldesmon is not related to is regulatory function, *FEBS Letters* 191:107(1985).

24. A. Husai-Chishti, A. Levin, and D. Branton, Abolition of actin bundling by phosphorylation of human erythrocyte protein 4.9, *Nature* 334:718(1988).

25. N.W. Ikebuchi and D.M.Waisman, Calcium-dependent regulation of actin filament bundling, *J. Biol. Chem.* 265:3392(1990).

26. M. Bahler and P. Greengard, Synapsin I bundles F-actin in a phosphorylation-dependent manner, *Nature* 326:704(1987).

27. K. Sobue, Y. Muramoto, M. Fujita, and S. Kakiuchi, Purification of a calmodulin-binding protein from chicken gizzard that interacts with F-actin, *Proc. Natl. Acad.Sci. USA* 78:5652(1981).

28. K. Sobue, K. Morimoto, K. Kanda, K. Maruyama, and S. Kakiuchi, Reconstitution of Ca^{2+}-sensitive gelation of actin filaments with filamin, caldesmon and calmodulin, *FEBS Letters* 138:289(1982).

29. E. Grazi, P. Cuneo, and A. Cataldi, The control of cellular shape and motility. Mg^{2+} and tropomyosin regulate the formation and the dissociation of microfilament bundles, *Biochem. J.* 288:727(1992).

30. A. Romani and A. Scarpa, Regulation of cell magnesium, *Archiv. Biochem. Biophys.* 298:1(1992).

31. A. Cittadini and A. Scarpa, Intracellular Mg^{2+} homeostasis of Ehrlich ascites tumor cells, *Archiv. Biochem. Biophys.* 227:202(1983).

32. E. Murphy, C.C. Freudenrich, L.A. Levy, and R.E. London, Monitoring cytosolic free magnesium in cultured chicken heart cells by use of the fluorescent indicator Furaptra, *Proc. Natl. Acad. Sci. USA* 86:2981(1989).

33. H. Holmsen, H.J. Day, and C.A. Setkowsky, Secretory mechanisms. Behaviour of adenine nucleotides during the platelet release reaction induced by adenosine diphosposphate and adrenaline, *Biochem. J.* 129:67(1972).

34. V. Pribluda, F. Laub, and A. Rotman, The state of actin in activated human platelets, *Eur. J. Biochem.* 116:293(1981).

35. D.E. Goll, A. Suzuki, J. Temple, and G.R. Holmes, Studies on purified alpha-actinin. Effect of temperature and tropomyosin on the alpha-actinin - F-actin interaction, *J. Mol. Biol.* 67:469(1972).

36. B.M. Jockusch and G. Isenberg, Interaction of alpha-actinin and vinculin with actin. Opposite effects on filament network formation, *Proc. Natl. Acad. Sci. USA* 78:3005(1981).

37. J.P. Bennett, K. Scott Zaner, and T.P. Stossel, Isolation and properties of macrophage alpha-actinin. Evidence that it is not an actin gelling protein, *Biochemistry* 23:5081(1984).

38. T. Ohtaki, S. Tsukita, N. Mimura, S. Tsukita, and A. Asano, Interaction of actinogelin with actin. No nucleation but high gelation activity, *Eur. J. Biochem.* 150:609(1985).

39. F. Landon, Y. Gache, H. Touitou, and A. Olomucki, Properties of two isoforms of human blood platelet alpha-actinin, *Eur. J. Biochem.* 153:231(1985).

40. E. Grazi, G. Trombetta, and M. Guidoboni, Microfilament gel rigidity cooperates negatively with the binding of actin gelling proteins, *Biochem. Int.* 21:633(1990).

41. E. Grazi, G. Trombetta, and M. Guidoboni, Binding of alpha-actinin to F-actin or to tropomyosin F-actin is a function of both alpha-actinin concentration and gel structure, *J. Muscle Res. Cell Mot.* 12:579(1991).

42. E. Grazi, G. Trombetta, E. Magri, and P. Cuneo, The actin gelling activity of chicken gizzard alha-actinin at physiological temperature is triggered by water sequestration, *FEBS Letters* 272:149(1990).

43. E. Grazi, P. Cuneo, E. Magri, and C. Schwienbacher, Preferential binding of alpha-actinin to actin bundles, *FEBS Letters* 314:348(1992).

44. E. Lazarides, Actin, alpha-actinin, and tropomyosin interaction in the structural organization of actin filaments in nonmuscle cells, *J. Cell Biol.* 68:202(1976).

45. T.D. Pollard and J.A. Cooper, Actin and actin-binding proteins. A critical evaluation of mechanisms and functions, *Ann. Rev. Biochem.* 55:987(1986).

46. P.A. Janmey and T.P. Stossel, Modulation of gelsolin function by phosphatidylinositol4,5-bisphosphate, *Nature* 325:362(1987).

47. E. Grazi, E. Magri, P. Cuneo, and A. Cataldi, The control of cellular motility and the role of gelsolin, *FEBS Letters* 295:163(1991).

48. J.M. Scholey, K.A. Taylor, and J. Kendrich Jones, Regulation of nonmuscle m y o s i n assembly by calmodulin-dependent light chain kinase, *Nature* 287:233(1980).

49. F.T. Ashton, A.V. Somlyo, and A.P. Somlyo, The contractile apparatus of vascular smooth muscle : Intermediate high voltage stereo electron microscopy, *J. Mol. Biol.* 98:17(1975).

50. A.V. Somlyo, T.M. Butler, M. Bond, and A.P. Somlyo, Myosin filaments have n o n - phosphorylated light chains in relaxed smooth muscle, *Nature* 294:567(1981).

51. E. Grazi and G. Trombetta, evidence that unphosphorylated smooth muscle myosin supports smooth muscle contraction, *Biochem. Biophys. Res. Communs.* 178:967(1991).

52. M. Barany, E. Polyak, and K. Barany, Protein phosphorylation during the contraction-relaxation-contraction cycle of arterial smooth muscle, *Archiv. Biochem. Biophys.* 294:71(1992).

53. A. Spacenko, J. Wagner, R. Dabrowska, and J.C. Ruegg, Caldesmon-induced inhibition of ATPase activity of actomyosin and contraction of skinned fibres of chicken gizzard smooth muscle, *FEBS Letters* 192:9(1985).

54. H. Katsuyama, C.L.A. Wan, and K.G. Morgan, Regulation of vascular smooth muscle tone by caldesmon, *J. Biol. Chem.* 267:14555(1992).

CELLULAR FUNCTIONS OF THE MICROFILAMENT SYSTEM

ACTIN FILAMENT DYNAMICS IN CELL MOTILITY

Julie A. Theriot

Department of Biochemistry and Biophysics
University of California, San Francisco
San Francisco, CA 94143-0448

INTRODUCTION

The motility of individual animal cells is important for a wide variety
of basic biological processes including rearrangements during embryonic
development, neurite outgrowth, wound healing, inflammation, and cancer
metastasis. The actin cytoskeleton in animal cells must also drastically
rearrange in a cell-cycle dependent manner in order to perform cytokinesis.
Actin microfilament dynamics are intimately involved in all of these common
types of cell motility, but the mechanisms of force transduction and movement
are still poorly understood.

The actin-based motility of different types of cells appears quite varied.
Amoebae move by extending pseudopods and then streaming their
cytoplasmic contents forward into the extensions. Many mammalian cells
move by extending actin-filled protrusions, thin finger-like filopodia or flat
veil-shaped lamellipodia, forming new attachments to their substrates, and
then contracting at the rear. Formation of such protrusions is not necessarily
associated with whole cell movement; platelets form impressive filopodia
upon activation to entangle themselves in clots, and growth cones at the tips of
growing neurites form filopodia and lamellipodia and move vigorously to
generate long axons without generally causing net translocation of the cell
body. Actin filament behavior and the control of that behavior is likely to be
quite different in pseudopods, filopodia and lamellipodia, and may well vary a
good deal even among protrusions of the same general type in different cells.

I will concentrate here on actin filament dynamics in lamellipodia.
Lamellipodia are formed by many types of moving epithelial cells, fibroblasts
and neurons. They are characteristically less than one micron thick but can be
up to ten microns across, and are filled with a dense isotropic meshwork of
actin filaments which excludes large organelles. Lamellipodia can rearrange
themselves very rapidly, totally changing shape in less than a minute in some
cells or persisting for many minutes in others. Although the chief function of

Actin: Biophysics, Biochemistry, and Cell Biology
Edited by J.E. Estes and P.J. Higgins, Plenum Press, New York, 1994

lamellipodia is to advance the leading edge of the cell forward, there appears to be a constant centripetal flux of material from the front edge of the lamellipodium backward toward the cell body in many types of motile cells (see for example Abercrombie et al., 1970a, b; Forscher and Smith, 1988; Fisher et al., 1988; Theriot and Mitchison 1992b). The relationships among actin filament dynamics and movements, centripetal flow of features on the dorsal surface of lamellipodia, and cell protrusion are not clear and are certainly complicated. Useful initial steps in understanding the process of lamellipodial protrusion would be to observe where and when actin filaments polymerize and depolymerize as the lamellipodium advances, and what other actin-associated proteins the cell uses to regulate and modulate filament dynamics.

First I will describe observations of actin filament behavior in cells relying mostly on the recently developed technique of photoactivation of fluorescence, and discuss how these observations have provided information on the localization of filament polymerization and depolymerization in lamellipodia of moving cells and in one related model system. I will then discuss what is currently understood about how one actin-binding protein, profilin, might be used by the cell to regulate actin filament polymerization in lamellipodia and in the model system.

IN VIVO OBSERVATIONS OF ACTIN FILAMENT DYNAMICS

The kinetic behavior of actin has been heavily studied *in vitro* using purified actin, but it is not immediately clear how our understanding of actin polymerization in a test tube relates to the real behavior of the cytoskeleton in the enormously complicated context of the cytoplasm of a living cell. Thus although the rate constants of monomer association and dissociation and the critical concentrations governing actin filament dynamics are known for a variety of *in vitro* conditions (Pollard, 1986; Drenckhahn and Pollard, 1986), it is not possible to realistically extrapolate the rates of polymerization and depolymerization of filaments in cells. It is necessary, then, to measure the relevant rates in cells as directly as possible.

Observations using fluorescence photoactivation in lamellipodia

The marking of cytoskeletal elements by fluorescence photoactivation has been used extensively to study the dynamic behavior of microtubules and actin in living cells (Mitchison, 1988; Theriot and Mitchison, 1991; Reinsch et al., 1991; Okabe and Hirokawa, 1992). The use of traditional fluorescent derivatives of cytoskeletal proteins *in vivo* can only reveal the steady state distribution of those proteins, and show changes in that steady state. Photoactivation, in contrast, can irreversibly mark a spatially defined subset of cytoskeletal elements. The movement and turnover of the marked filaments can then be followed over time using fluorescence videomicroscopy.

For studies of actin filament dynamics, a derivative of the red fluorescent compound resorufin, caged resorufin iodoacetamide (CR iodoacetamide) was synthesized and covalently coupled to purified rabbit skeletal muscle actin through cysteine 374, near the C terminus (Figure 1; see Theriot and Mitchison 1991 for the synthesis and properties of CR iodoacetamide). CR-actin readily copolymerizes with unlabeled actin *in vitro*, and will incorporate into endogenous actin structures including stress fibers and lamellipodia when injected into cells (Theriot and Mitchison, 1991). CR-actin is stably non-fluorescent, but is converted to bright red fluorescent resorufin-actin upon exposure to 360-nm ultraviolet light (Figure 1).

Actin filament dynamics in lamellipodia of three different motile cell types, goldfish epidermal keratocytes (Theriot and Mitchison, 1991), IMR 90 human lung fibroblasts, and MC7 3T3 embryonic mouse fibroblasts (Theriot and Mitchison, 1992b) were examined using CR-actin. In each case, CR-actin was injected into the cells and allowed to equilibrate with endogenous actin pools for 30 minutes to 1 hour. Then a slit-shaped microbeam of UV light (Mitchison, 1988) was used to photoactivate a narrow bar of actin filaments in

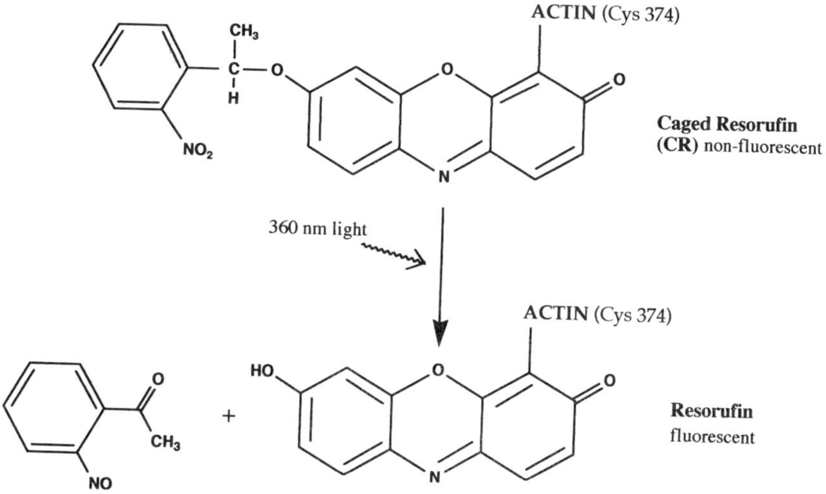

Figure 1. Structure of caged resorufin-actin. Actin covalently coupled to CR through cysteine 374 is irreversibly converted to highly fluorescent resorufin-actin upon exposure to 360 nm ultraviolet light.

the lamellipodia, parallel to the leading edge and therefore perpendicular to the direction of movement.

In both types of fibroblast, the actin filaments in the activated bar were observed to move with respect to the substrate centripetally toward the cell body (Theriot and Mitchison, 1992b). This result is consistent with earlier observations using photobleaching of labeled actin in a third fibroblast type (Wang, 1985) and in neuron growth cones (Okabe and Hirokawa, 1991). However, the rate of centripetal flux of actin relative to the substrate was in no way correlated with the rate of lamellipodial protrusion in any cell. In keratocytes, the activated actin bars remained basically stationary with respect to the substrate as the cells moved forward, regardless of how rapidly the cells were moving (Theriot and Mitchison, 1991). The rates are summarized in Table I. In no cases were the activated filaments observed to move forward as the lamellipodium protruded. Thus as the lamellipodium advances, the meshwork of actin filaments filling the new volume must arise entirely through new polymerization, rather than through forward movement of preexisting filaments (see below, also Figure 3). Furthermore, since the rate of centripetal flux of actin filaments relative to the substrate is independent of the rate of lamellipodial protrusion for all cell types observed, the control

Table I. Dynamic parameters of actin-based motility in keratocyte and fibroblast lamellipodia and in *L. monocytogenes* comet tails.

	Keratocyte[1]	IMR 90[2]	MC7 3T3[2]	*L. monocytogenes*[3]
rate of protrusion or propulsion (μm/min)	2.40 (1.74)[4]	0.41 (0.55)	0.08 (0.20)	4.38 (2.76)
rate of actin movement (μm/min)	-0.11 (0.70)	-0.66 (0.27)	-0.36 (0.16)	- 0 -
actin half-life (sec)	23 (6)	55 (28)	181 (99)	33 (16)
half-lives to cross structure	10	8	5	approx. 2

[1]Data for keratocytes (fish epidermal cells) are from Theriot and Mitchison, 1991.
[2]Data for IMR 90 human fibroblasts and MC7 3T3 mouse fibroblasts are from Theriot and Mitchison, 1992a.
[3]Data for *L. monocytogenes* are from Theriot et al., 1992.
[4]Standard deviations for all measurements are given in parentheses.

mechanisms governing the rate of protrusion must operate solely at the leading edge, and the behavior of actin filaments throughout the rest of the lamellipodium must be fairly independent of the rate of new actin insertion at the leading edge.

The average half-life of actin filaments in the lamellipodia was measured by observing the rate of decay of the fluorescent signal arising from photoactivation. The stability of the actin filaments in the lamellipodial meshwork appears to be inversely correlated with the rate of lamellipodial protrusion (Table I). In each cell type, the rate of filament disappearance remained fairly constant over the entire breadth of the lamellipodium.

Determination of sites of polymerization and depolymerization in lamellipodia

Since there was no correlation between the average filament half-life and the position of the activated bar in the lamellipodium, we can infer that actin filaments depolymerize at approximately the same rate and with approximately the same probability everywhere in the lamellipodium. In other words, the control of filament depolymerization is not spatially localized in the bulk of the lamellipodium. However, at the back end of the lamellipodium nearest the cell body, there is a substantial and precipitate drop-off in actin filament density (Figure 2). At this point in space there must be some biochemical singularity, which drastically accelerates the rate of filament depolymerization, to destroy the lamellipodial meshwork. The mechanism by which the cell accomplishes this is presently unknown.

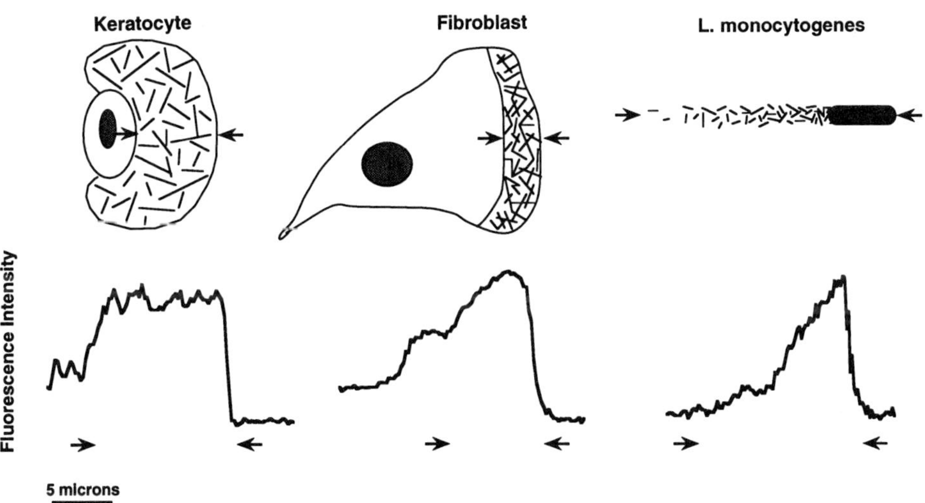

Figure 2. Actin filament density distribution in lamellipodia of keratocytes and fibroblasts, and in *L. monocytogenes* comet tails. Traces show fluorescence intensity profiles in representative cells which have been fixed and labeled with rhodamine-phalloidin. Arrows indicate planes of section and boundaries of structures.

By comparing the dynamic observations using photoactivation with static observations of steady state actin filament distribution in lamellipodia, it is possible also to infer the location of filament polymerization. In all lamellipodia actin filaments move backwards relative to the leading edge of the cell, and an actin filament born at the leading edge will eventually reach the rear of the lamellipodium. In the very slowly moving MC7 3T3 cells, the filaments will cross the lamellipodium almost entirely via centripetal flux relative to the substrate. In keratocytes, the filaments will cross the lamellipodium almost entirely by virtue of having the lamellipodium move forward over them as they remain fixed in space. In IMR 90 cells, both processes contribute to the movement of filaments relative to the leading edge. Having measured the rate of lamellipodial protrusion, the rate of actin filament movement relative to the substrate, and the average size of lamellipodia in each cell type, it is possible to calculate the length of time it would take for a filament born at the leading edge to reach the back of the lamellipodium. It is most relevant to express this time as a multiple of the average filament half-life in each cell type (Table I). In all three types of lamellipodia, it takes several average half-lives (between 5 and 10) to cross.

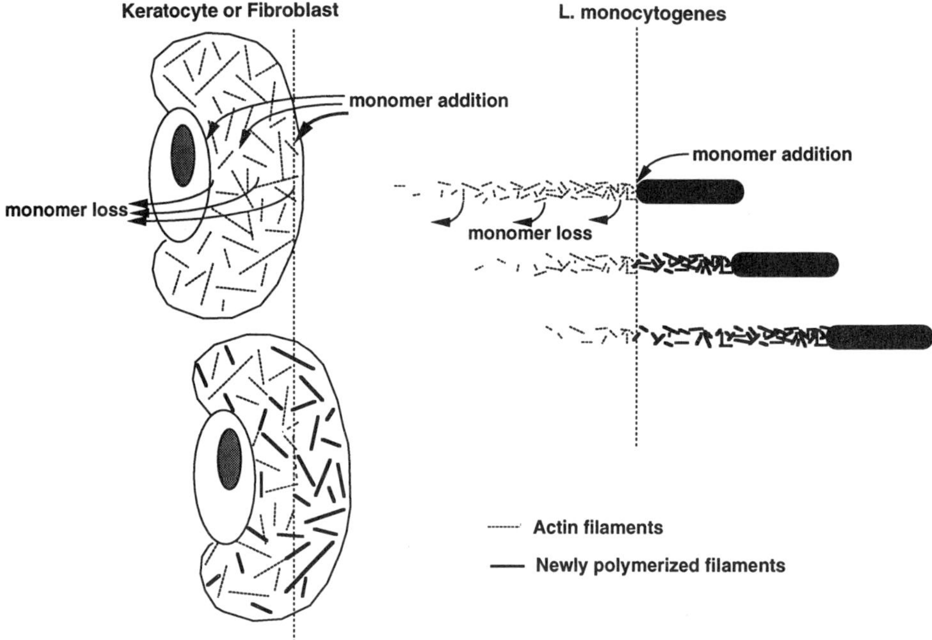

Figure 3. Sites of polymerization and depolymerization of actin filaments in lamellipodia and *L. monocytogenes* tails. In lamellipodia, polymerization occurs everywhere but is most rapid at the leading edge. In comet tails, polymerization occurs only at the front. In both cases, depolymerization occurs uniformly throughout the structure.

Some models of actin filament dynamics in lamellipodia favor the idea that filaments assemble only right at the leading edge (for example, Wang, 1985). If filaments were polymerized only at the leading edge but depolymerized uniformly elsewhere, the relatively rapid turnover times of actin in the lamellipodia would create a steep gradient of filament density distribution. For example, in MC7 3T3 cells, it would take 5 half-lives for the filaments to cross the lamellipodium and only $1/2^5$ or about 3% of the filaments seen at the leading edge should persist at the back of the lamellipodium. This is not the case; there is a shallow gradient of filament distribution in MC7 3T3 fibroblast lamellipodia (Figure 2), but the density falls off on average only about 40%, not the 97% that such a model would predict. For keratocytes, the situation is even more extreme; it would take approximately 10 average half-lives for a filament to cross the lamellipodium, but the density distribution is essentially flat (Figure 2). Thus in all cell types, there must be some actin polymerization everywhere throughout the lamellipodium, as well as depolymerization everywhere, to maintain the steady-state distribution observed (Figure 3; see also Theriot and Mitchison 1992a for a more thorough discussion of this issue). In rapidly advancing lamellipodia, the amount of filament polymerization at the leading edge must still be substantially faster than elsewhere in the lamellipodium, to create the volume of new actin meshwork needed at the advancing edge.

Intracellular propulsion of *Listeria monocytogenes*

Describing the spatial and temporal changes in actin filament dynamics and movements is only one initial step in understanding the role of actin in cell motility. Ultimately it will be necessary to understand the mechanisms by which the cell regulates actin dynamics to achieve coordinated movement. For example, it would be particularly useful to identify the factors which are responsible for the precipitate loss of actin filaments at the rear of the lamellipodium, and which promote actin filament polymerization throughout the lamellipodium but more strongly at the leading edge. Genetic and reverse-genetic approaches to dissecting the control of cell motility have begun to provide useful insight into the role of particular actin-binding proteins in whole-cell motility (see for example DeLozanne and Spudich, 1987; Knecht and Loomis, 1987; Cox et al., 1992). However, biochemical dissection of the mechanisms of lamellipodial protrusion seems to be particularly daunting. The chief obstacle to reconstituting lamellipodial movements in a simplified cell-free system is the intimate role of the plasma membrane in localizing actin filament nucleation (see for example, Shariff and Luna, 1992) and perhaps also in generating force (Oster and Perelson, 1987). It would obviously be highly desirable to find a model system which mimics the actin filament dynamics involved in lamellipodial protrusion in the absence of the plasma membrane.

One model system that appears to mimic lamellipodial protrusion well is the actin-based propulsion of the intracellular bacterial parasite *Listeria monocytogenes*. *L. monocytogenes* is the causative agent of listeriosis, a serious food-borne infection that can cause miscarriages in healthy pregnant women and meningitis in immunocompromised people (Gellin and Broome, 1989). When the bacteria come into contact with the surface of a host epithelial cell, they induce their own phagocytosis. Once inside an endosome they secrete enzymes which degrade the endosomal membrane and are released free directly in the host cell cytoplasm, where they begin to grow and divide. Within a few hours after infection, host cell actin filaments initially form a dense cloud around the intracytoplasmic bacteria, and then rearrange to form a polarized "comet tail." The comet tail is made up of very short (0.2 µm) actin filaments crosslinked into a random meshwork in which the majority of filaments have their barbed (rapidly growing) ends oriented toward the bacterium (Tilney and Portnoy, 1989). This is quite reminiscent of the organization of the lamellipodium, which also appears to be a loose mesh with some preference for barbed-end forward orientation (Small, 1981; Lewis and Bridgman, 1993). Bacteria with comet tails move very rapidly through the host cell cytoplasm (up to 1.4 µm/sec in some types of tissue culture cells; Dabiri et al., 1990) and push out long, membrane-bound protrusions that mediate the spread of infection between neighboring cells.

Experiments involving microinjection of fluorescently labeled actin indicated that new actin filaments polymerize primarily at or near the surface of the bacterium (Sanger et al., 1992). Photoactivation experiments showed that the actin filaments of the tail always remain stationary in the cytoplasm, regardless of how rapidly the bacteria move (Theriot et al., 1992). Thus polymerization at the front of the comet tail occurs at the same rate as bacterial movement and must entirely account for the remarkably rapid rates of propulsion observed. This is analogous to the situation in keratocyte lamellipodia, where the actin filament meshwork remains stationary with respect to the substrate regardless of the rate of lamellipodial protrusion (Theriot and Mitchison, 1991; see Table I). Quantitative analysis of the photoactivation observations and comparison of these dynamic observations with the steady-state filament distribution revealed several other interesting results. In comet tails, as in lamellipodia, depolymerization occurs uniformly everywhere. Unlike in lamellipodia, however, filament polymerization occurs *only* at the front, immediately adjacent to the bacterial surface. Thus an exponential drop-off of filament density is observed (Figure 2), and it takes only about 2 half-lives for a filament born at the front of the comet tail, next to the bacterium, to traverse most of the tail (Table I).

The *L. monocytogenes* comet tail, then, behaves like a slightly simplified lamellipodium (Theriot et al., 1992). There is no centripetal flux of actin filaments relative to the stationary substrate, so a major complication characteristic of fibroblast lamellipodia is removed. Furthermore, the site of filament polymerization is very tightly localized, occuring exclusively at the front of the tail near the bacterium (Figure 3). Since intracellular *L. monocytogenes* propulsion does not directly involve the host cell plasma membrane, it ought theoretically be possible to reconstitute this type of actin-based motility in a cell-free cytoplasmic extract. In fact, such a reconstitution has been achieved, opening the door to a traditional biochemical approach to dissect the control of filament polymerization and depolymerization in this motile system (MS in preparation).

MODULATION OF ACTIN FILAMENT DYNAMICS BY ACTIN-BINDING PROTEINS

Armed with these descriptions of actin filament behavior during cell motility, I would now like to shift focus to a discussion of how the cell might use certain known actin-associated proteins to modulate actin dynamics in lamellipodia and in the analogous *L. monocytogenes* comet tails. Again the challenge to the field is to successfully apply the wealth of available information regarding the effects of numerous known actin-associated proteins on the dynamics of pure actin *in vitro* to the complex and unpredictable environment of the cell.

Control of filament polymerization: nucleation vs. desequestration

In lamellipodia and in *L. monocytogenes* comet tails, actin filament polymerization is spatially regulated. Most animal cells contain large pools of unpolymerized actin; while the total concentration of actin in the cytoplasm is often in excess of 100 μM, only about 50% of the actin in most cells is found in the insoluble filamentous cytoskeleton (Blikstad et al, 1978). At such high concentrations, purified actin *in vitro* would be >99% in filaments (Pollard 1986). Thus the cell must have regulatory mechanisms which actively discourage polymerization, and the leading edges of lamellipodia and the surfaces of motile intracellular *L. monocytogenes* must probably use specific mechanisms to overcome this general inhibition of polymerization.

There are in theory at least two ways the cell might limit the extent of actin polymerization. First, the rate-limiting step in actin polymerization is the formation of nuclei; monomers are more efficient at adding onto the ends of preexisting filaments. Thus the cell might discourage actin polymerization by limiting the number of available nuclei, for example by keeping the ends of existing filaments capped and therefore unavailable for monomer addition. This seems quite plausible, since there are many known actin filament-binding proteins which act as caps on barbed ends. Alternatively, the cell might maintain its large pool of unpolymerized actin by sequestering most of the actin monomers, keeping them in a state which will not readily polymerize. For example, most of the monomeric actin in the cell might be bound in a complex with another protein which does not allow the actin to add onto a filament. This too is plausible, since a variety of such sequestering proteins have been identified and one (thymosin β_4; Safer et al., 1990) is present in most cells in sufficiently large amounts to nearly account for the amount of unpolymerized actin observed.

One recent experiment showed quite decisively that monomer sequestration is largely responsible for the limitation of actin polymerization. Phalloidin-stabilized actin filaments, which should act as excellent nuclei for elongation of filaments, do not induce noticeable polymerization when injected into cells (Sanders and Wang, 1990). This failure must be due to sequestration of monomer rather than to rapid capping, since purified exogenous actin injected along with or shortly after the nuclei readily elongates them. The nuclei do get capped within about an hour after being injected, so both mechanisms do operate to some extent, but monomer sequestration appears to be the chief one. Thus in order for the leading edges of lamellipodia or *L. monocytogenes* to induce polymerization in a spatially regulated manner, actin monomers must be locally desequestered.

Monomer desequestration via nucleotide exchange

The most obvious way desequestration could be spatially regulated would be for some protein to specifically disrupt the bond between actin monomer and thymosin β_4. As yet, no protein is known to have this function. Another possibility is that actin monomers might be effectively desequestered by factors which modulate the binding of adenine nucleotides to actin.

Profilin is an actin monomer-binding protein which can sequester monomers when bound at a 1:1 molar ratio. The binding of profilin to actin monomer increases the rate constant for dissociation of nucleotide by three orders of magnitude (Goldschmidt-Clermont et al., 1991). More and more known GTPases are being found to have associated guanine nucleotide exchange factors which play important roles in their regulation; by analogy, profilin might act primarily in most cells as an adenine nucleotide exchange factor for the actin ATPase.

Thymosin β_4, in contrast, strongly inhibits nucleotide exchange on actin (Goldschmidt-Clermont et al., 1992a). Interestingly, in a mixture of actin, thymosin β_4, and profilin, both the thymosin β_4 and the profilin come on and off actin rapidly enough that profilin can dramatically increase nucleotide exchange even which it is present at very low molar ratios (Goldschmidt-Clermont et al., 1992a). Thus spatial and temporal regulation of profilin concentration might have profound effects on the rate of nucleotide exchange on actin monomer.

The way in which this effect could regulate actin filament polymerization is diagrammed in Figure 4. As old actin filaments depolymerize, ADP-actin monomers are released and quickly bound up by thymosin β_4. As long as the concentration of free ADP-actin is kept buffered below its critical concentration, about 2 μM (Pollard 1986), no spontaneous polymerization will occur, and the binding to thymosin β_4 will make spontaneous nucleotide exchange very slow. However, in the presence of profilin, some ADP-actin will be converted to ATP-actin. ATP-actin polymerizes somewhat more rapidly than ADP-actin. More importantly, its critical concentration for polymerization is about 10-fold lower. Thus in a region of the cell where the ubiquitous pool of about 2 μM ADP-actin is suddenly converted to ATP-actin, polymerization will readily occur.

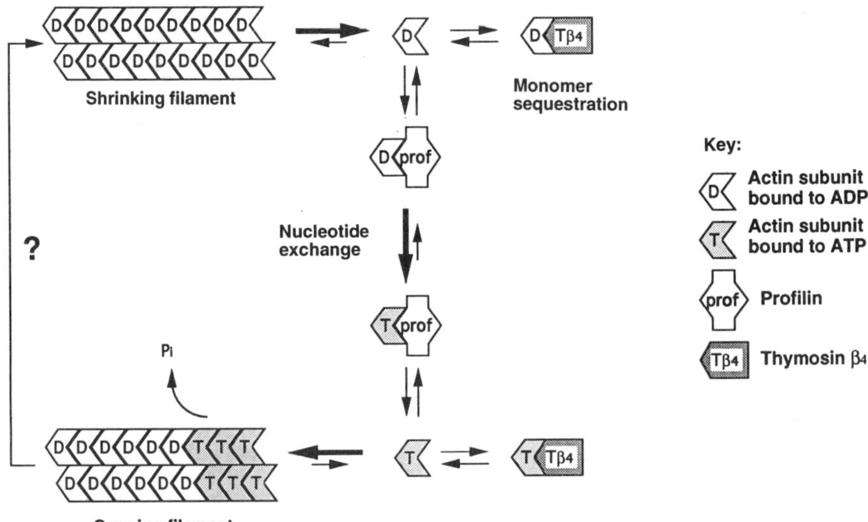

Figure 4. Monomer desequestration through nucleotide exchange. As filaments depolymerize, ADP-bound actin monomers are released and are rapidly bound up by thymosin β_4 in the cytoplasm, which prevents nucleotide exchange and buffers the concentration of free ADP-actin below its critical concentration of about 2 μM. In the presence of profilin, some ADP-actin is converted to ATP-actin. Since ATP actin polymerizes at a critical concentration 10-fold lower than ADP-actin, filament elongation can occur.

Possible role of profilin in cell motility

Actin filament polymerization occurs everywhere throughout lamellipodia, and more rapidly at the leading edge during rapid protrusion. Polymerization occurs only near the bacterial surface in *L. monocytogenes* comet tails (see above, and Figure 3). Thus we would expect the actin-associated proteins which are responsible for promoting polymerization in these two motile systems to be located throughout lamellipodia, and on the surface of motile intracellular *L. monocytogenes*. Profilin fulfills these predictions; by immunolocalization, it is found highly concentrated throughout lamellipodia (as well as present on membranes elsewhere in the cell body) (Buß et al., 1992), and is specifically localized on the back end of moving intracellular *L. monocytogenes*, where it is probably bound to a bacterial surface protein known to be necessary for motility (MS in preparation).

Although this is suggestive, it is merely a correlation, and it is always difficult to extend information about the *in vitro* effects of actin binding proteins on filament dynamics to the cytoskeleton of a living cell. There is now, however, some *in vivo* evidence that a major function of profilin inside cells is to stabilize actin filaments. In CHO (Chinese Hamster Ovary) cells stably overexpressing human profilin, the average actin filament half-life as measured by decay of photoactivated fluorescence is directly proportional to the amount of profilin present; cells with more profilin contain more stable actin filaments. Furthermore, cells overexpressing profilin are more resistant to the effects of cytochalasin D (Goldschmidt-Clermont et al., 1992b). This increase in stability is consistent with the scheme outlined in Figure 4 if we assume that the increased amount of ATP-actin available allows individual filaments to grow for a longer time before they begin to shrink (see Theriot and Mitchison, 1992a). Microinjection of profilin, on the other hand, causes a decrease in the number of actin filaments in tissue culture cells, although the stability of the remaining filaments has not been examined (Cao et al., 1992). It is difficult to directly compare the results of experiments involving transient and disruptive microinjection to stable overexpression, and more work is needed to resolve this issue. At present, though, the possibility that profilin may generally promote actin filament polymerization in most cells and might serve to spatially localize polymerization during motility remains viable.

CONCLUSIONS

As more becomes known about actin and actin-binding proteins at an increasing detailed biochemical and structural level, we can begin to dissect the complex dynamic behavior of the actin cytoskeleton as a whole during such processes as cell motility. Development and application of techniques to more or less directly observe cytoskeletal dynamics in living cells, such as fluorescence photoactivation, will remain important. Comparative studies on lamellipodia of different cell types and on *L. monocytogenes* comet tails have revealed unifying themes on how actin dynamics are regulated during motility, and will continue to direct and feed back on studies of the roles of individual known actin-binding proteins in cell motility.

REFERENCES

Abercrombie, M., Heaysman, J. E. M., and Pegrum, S. M., 1970a, The locomotion of fibroblasts in culture. II. "Ruffling." *Exp. Cell Res.* 59: 393-398.

Abercrombie, M., Heaysman, J. E. M., and Pegrum, S. M., 1970b, The locomotion of fibroblasts in culture. III. Movements of particles on the dorsal surface of the leading lamella. *Exp. Cell Res.* 62: 389-398.

Blikstad, I., Markey, F., Carlsson, L., Persson, T., and Lindberg, U., 1978, Selective assay of monomeric and filamentous actin in cell extracts, using inhibition of deoxyribonuclease I, *Cell* 15: 935-943.

Buβ, F., Temm-Grove, C., Henning, S., and Jockusch, B. M., 1992, Distribution of profilin in fibroblasts correlates with the presence of highly dynamic actin filaments, *Cell Motil. Cytoskel.* 22: 51-61.

Cao, L.-g., Babcock, G. G., Rubenstein, P. A., and Wang, Y.-l., 1992, Effects of profilin and profilactin on actin structure and function in living cells, *J. Cell Biol.* 117: 1023-1029.

Cox, D., Condeelis, J., Wessels, D., Soll, D., Kern, H., and Knecht, D. A., 1992, Targeted disruption of the ABP-120 gene leads to cells with altered motility, *J. Cell Biol.* 116: 943-955.

Dabiri, G. A., Sanger, J. M., Portnoy, D. A. and Southwick, F. S., 1990, *Listeria monocytogenes* moves rapidly through the host-cell cytoplasm by inducing directional actin assembly, *Proc. Natl. Acad. Sci. USA* 87: 6068-6072.

DeLozanne, A., and Spudich, J. A., 1987, Disruption of the *Dictyostelium* myosin heavy chain gene by homologous recombination,*Science* 236: 1086-1091.

Drenckhahn, D., and Pollard, T. D., 1986, Elongation of actin filaments is a diffusion-limited reaction at the barbed end and is accelerated by inert macromolecules, *J. Biol. Chem.* 261: 12754-12758.

Fisher, G. W., Conrad, P. A., DeBiasio, R. L., and Taylor, D. L., 1988, Centripetal transport of cytoplasm, actin and the cell surface in lamellipodia of fibroblasts, *Cell Motil. Cytoskel.* 11:235-247.

Forscher, P., and Smith, S. J., 1988, Actions of cytochalasins on the organization of actin filaments and microtubules in a neuronal growth cone, *J. Cell Biol.* 107: 1505-1516.

Gellin, B. G., and Broome, C. V., 1989, Listeriosis, *JAMA* 261:1313-1320.

Goldschmidt-Clermont, P. J., Machesky, L. M., Doberstein, S. K., and Pollard, T. D., 1991, Mechanism of the interaction of platelet profilin with actin, *J. Cell Biol.* 113: 1081-1089.

Goldschmidt-Clermont, P. J., Furman, M. I., Wachsstock, D., Safer, D., Nachmias, V. T., and Pollard, T. D., 1992a, The control of actin nucleotide exchange by thymosin β4 and profilin. A potential regulatory mechanism for actin polymerization in cells, *Mol. Biol. Cell* 3: 1015-1024.

Goldschmidt-Clermont, P. J., Theriot, J. A., Tomaselli, G. F., and Finkel, T., 1992b, Profilin overexpression stabilizes actin filament bundles, *Circulation* 86: I-178.

Knecht, D. A., and Loomis, W. F., 1987, Antisense RNA inactivation of myosin heavy chain gene expression in *Dictyostelium discoideum*, *Science* 236: 1081-1085.

Lewis, A. K., and Bridgman, P. C., 1992, Nerve growth cone lamellipodia contain two populations of actin filaments that differ in organization and polarity, *J. Cell Biol.* 119: 1219-1243.

Mitchison, T. J., 1988, Polewards microtubule flux in the mitotic spindle: evidence from photoactivation of fluorescence, *J. Cell Biol.* 109: 637-652.

Okabe, S., and Hirokawa, N., 1991, Actin dynamics in growth cones, *J. Neurosci.* 11: 1918-1929.

Okabe, S. and Hirokawa, N., 1992, Differential behavior of photoactivated microtubules in growing axons of mouse and frog neurons, *J. Cell Biol.* 117: 105-120.

Oster, G. F., and Perelson, A. S., 1987, The physics of cell motility, *J. Cell Sci. Suppl.* 8: 35-54.

Pollard, T. D., 1990, Rate constants for the reactions of ATP- and ADP-actin with the ends of actin filaments, *J. Cell Biol.* 103: 2747-2754.

Reinsch, S. S., Mitchison, T. J., and Kirschner, M., 1991, Microtubule polymer assembly and transport during axonal elongation, *J. Cell Biol.* 115: 365-379.

Safer, D., Golla, R., and Nachmias, V. T., 1990, Isolation of a 5-kilodalton actin-sequestering peptide from human blood platelets, *Proc. Natl. Acad. Sci. USA* 87: 2536-2540.

Sanders, M. C., and Wang, Y.-l., 1990, Exogenous nucleation sites fail to induce detectable polymerization of actin in living cells, *J. Cell Biol.* 110: 359-365.

Sanger, J. M., Sanger, J. W., and Southwick, F. S., 1992, Host cell actin assembly is necessary and likely to provide the propulsive force for intracellular movement of *Listeria monocytogenes, Infect. Immun.* 60: 3609-3619.

Shariff, A., and Luna, E. J., 1992, Diacylglycerol-stimulated formation of actin nucleation sites at plasma membranes, *Science* 256: 245-247.

Small, J. V., 1981, Organization of actin in the leading edge of cultured cells: influence of osmium tetroxide and dehydration on the ultrastructure of actin meshworks, *J. Cell Biol.* 91: 695-705.

Theriot, J. A., and Mitchison, T. J., 1991, Actin microfilament dynamics in locomoting cells, *Nature* 352: 126-131.

Theriot, J. A., and Mitchison, T. J., 1992a, The nucleation-release model of actin filament dynamics in cell motility, *Trends Cell Biol.* 2: 219-222.

Theriot, J. A., and Mitchison, T. J., 1992b, Comparison of actin and cell surface dynamics in motile fibroblasts, *J. Cell Biol.* 118: 367-377.

Theriot, J. A., Mitchison, T. J., Tilney, L. G., and Portnoy, D. A., 1992, The rate of actin-based motility of intracellular Listeria monocytogenes equals the rate of actin polymerization, *Nature* 357: 257-260.

Tilney, L. G., and Portnoy, D. A., 1989, Actin filaments and the growth, movement, and spread of the intracellular bacterial parasite, Listeria monocytogenes, *J. Cell Biol.* 109: 1597-1608.

Wang, Y.-l., 1985, Exchange of actin subunits at the leading edge of living fibroblasts: possible role of treadmilling, *J. Cell Biol.* 101: 597-602.

CHANGES IN ADHESION PLAQUE PROTEIN LEVELS REGULATE CELL MOTILITY AND TUMORIGENICITY

Avri Ben-Ze'ev,[1] José Luis Rodríguez Fernández[1], Ursula Glück[1], Daniela Salomon[2] and Benjamin Geiger[2]

[1]Department of Molecular Genetics and Virology
[2]Chemical Immunology
Weizmann Institute of Science
Rehovot, 76100, Israel

INTRODUCTION

Cell adhesion to neighboring cells and to the extracellular matrix (ECM) plays a major role in cell and tissue morphogenesis (Edelman, 1992; Takeichi, 1991; Hynes, 1992). These complex, adhesion-related cellular processes are mediated through transmembrane contact receptors of the cadherin and integrin families of receptors (Takeichi, 1991; Hynes, 1992). In the cytoplasmic domain, these receptors interact with cytoskeletal plaque proteins such as vinculin, talin and α-actinin which anchor the microfilament system to junctional areas in adherens type junctions (AJ) in adhesion plaques, and to α and β catenin and plakoglobin in cell-cell AJ (Burridge et al., 1988; Geiger and Ginsberg, 1991; Geiger et al., 1992). The cascade of molecular interactions which links the outside to the inside of the cell defines cell shape and motility, and also has a function in signal transduction which results in effects on cell growth, differentiation, and gene expression (Ben-Ze'ev, 1991; 1992; Schwartz, 1992; Haskill and Juliano, 1993). Signaling through adhesion plaques is suggested to occur through changes in tyrosine phosphorylation (Burridge et al., 1992; Volberg et al., 1992). Moreover, recent studies have demonstrated that the changes in tyrosine phosphorylation of a cytoplasmic adhesion plaque tyrosine kinase (p125^{FAK}) is common to adhesion related signaling and to growth factor, cytokine and neuropeptide induced signaling (Zachary and Rozengurt, 1992), and that tyrosine phosphorylation of p125^{FAK} is constitutively activated in oncogene-transformed cells (Guan and Shalloway, 1992). These results suggest a convergence, in adhesion plaques, of signals transduced by cytokines, oncogenes and adhesion.

In addition to changes in the organization of AJ proteins by posttranslational modifications, modulations in the expression of cytoskeletal plaque proteins were described in cells stimulated to locomote, proliferate, differentiate, and in oncogene-transformation (Ben-Ze'ev, 1985; 1992; Raz and Ben-Ze'ev, 1987). We describe here that such changes in individual AJ protein levels can have a profound effect on cell motility, growth and tumorigenic ability of cells, and may serve as physiological means for controlling cell behavior.

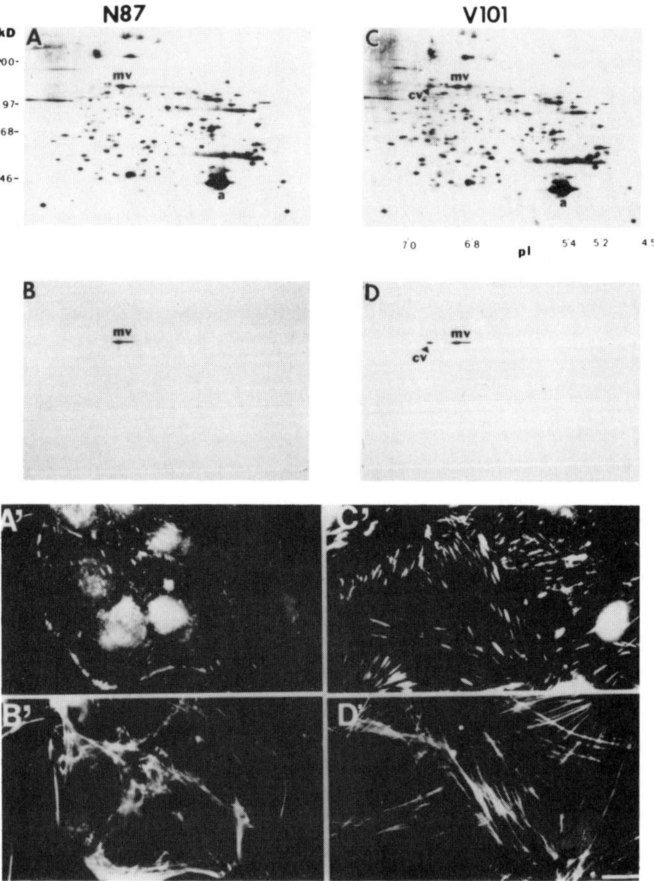

Figure 1. Levels of chicken vinculin and its organization in transfected 3T3 cells. Cells of a neo^r clone N87 (A, B, A', B') and of a chicken vinculin-transfected clone (C, D, C', D') V101 were labeled for 16 hours with ^{35}S-methionine (A-D), or processed for double immunofluorescence for vinculin (A', B') and actin (C', D'). Equal amounts of radioactive total cell protein were analyzed by 2-D gel electrophoresis, transferred to nitrocellulose and the protein pattern was obtained by autoradiography (A, C). The same blots were reacted with a broad-species type anti vinculin antibody, followed by alkaline phosphatase-coupled anti mouse IgG (B, D). **mv**, mouse vinculin; **cv**, chicken vinculin; **a**, actin.

OVEREXPRESSION OF VINCULIN INHIBITS THE MOTILITY OF 3T3 CELLS

We have chosen to selectively modulate vinculin expression in 3T3 cells, since previous studies have shown that the expression of vinculin, a major component of adhesion plaques and cell-cell AJ is modulated under a variety of physiological conditions. For example, vinculin RNA and protein levels are regulated in fibroblasts establishing contact on different ECMs (Ungar et al., 1986; Bendori et al., 1987), in differentiating adipocytes (Rodríguez Fernández and Ben-Ze'ev, 1989) and granulosa cells (Ben-Ze'ev and Amsterdam, 1987), in cultured smooth muscle cells (Belkin et al., 1988), and in migrating corneal epithelial cells (Zieske et al.,1989). Furthermore, recent studies have shown that expression of the vinculin gene and protein are rapidly yet transiently induced in growth stimulated 3T3 cells (Ben-Ze'ev et al., 1990; Bellas et al., 1991), and in regenerating liver following partial hepatectomy (Glück et al., 1992). This transcriptional activation occurs, most probably, by activating a serum response element localized in the vinculin promoter region (Moiseyeva et al., 1993).

The results summarized in Figures 1 and 2 demonstrate that in 3T3 cells stably overexpressing chicken vinculin by only 20% above the endogenous protein level (Figure 1A-D), there is an increase in the number and size of stress fibers and vinculin-containing adhesion plaques (Figure 1C', 1D', compare to control Figure 1A', 1B'). Moreover, the motility of such cells was dramatically reduced when measuring the ability of the cells to

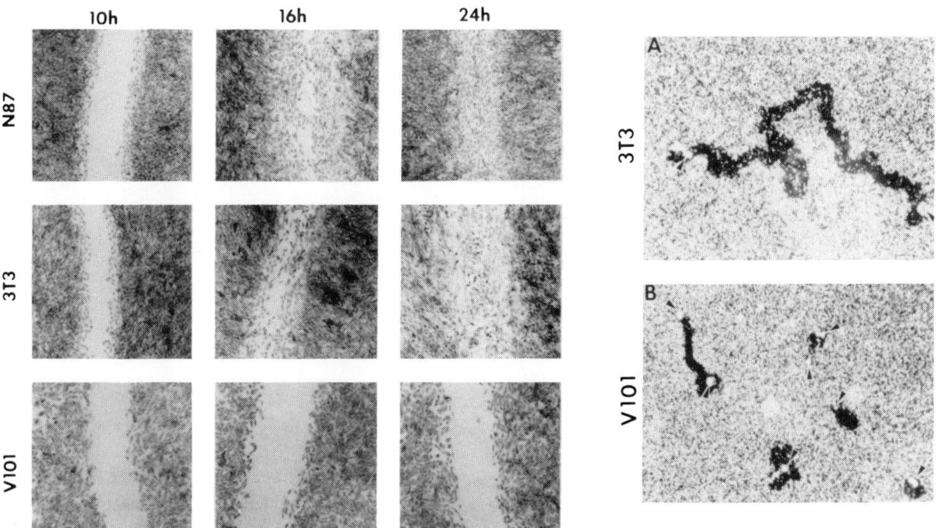

Figure 2. Reduced motility of vinculin-transfected cells. 3T3 cells, transfected cells with the neo^r gene alone (N87), and cells transfected with chicken vinculin (V101), were cultured to confluence. A wound was introduced in the monolayers and the cultures were incubated in fresh medium. At 10, 16, and 24 hours after the "wounding", the cultures were fixed and stained with Giemsa. (A, B) 500 cells were seeded per 20x20 mm coverslip precoated with colloidal gold. The phagokinetic tracks produced by 3T3 cells (A) were compared, 24 hours after seeding the cells, to those of vinculin-transfected cells (clone V101) by darkfield microscopy (B) (From Rodríguez Fernández et al., 1992a).

close a "wound" introduced in a confluent monolayer (Figure 2, V101), and by determining the length of phagokinetic tracks produced by the cells on colloidal gold-coated dishes (Figure 2, V101, compare to 3T3 and the neo^r N87 control clone). We were unable to obtain stable 3T3 clones expressing higher levels of the transfected vinculin using different expression vectors. The results suggest that a moderate elevation in vinculin content can have a significant inhibitory effect on the motility of 3T3 cells.

SUPPRESSION OF VINCULIN EXPRESSION IN 3T3 CELLS EFFECTS CELL SHAPE, MOTILITY AND ANCHORAGE DEPENDENCE

To study the effect of reduced vinculin levels on cell behavior, an antisense vinculin expression vector containing part of the mouse vinculin cDNA was utilized (Rodríguez Fernández et al., 1993). 3T3 clones co-transfected with the neo^r gene and the vinculin antisense cDNA were screened by quantitative Western blotting to identify stable clones expressing decreased vinculin levels. Clones expressing suppressed vinculin levels down to between 10% to 30% of control 3T3 cells were obtained (Figure 3 I). These clones had a round morphology (Figure 3C, E and F, compare to control, Figure 3D), and only few

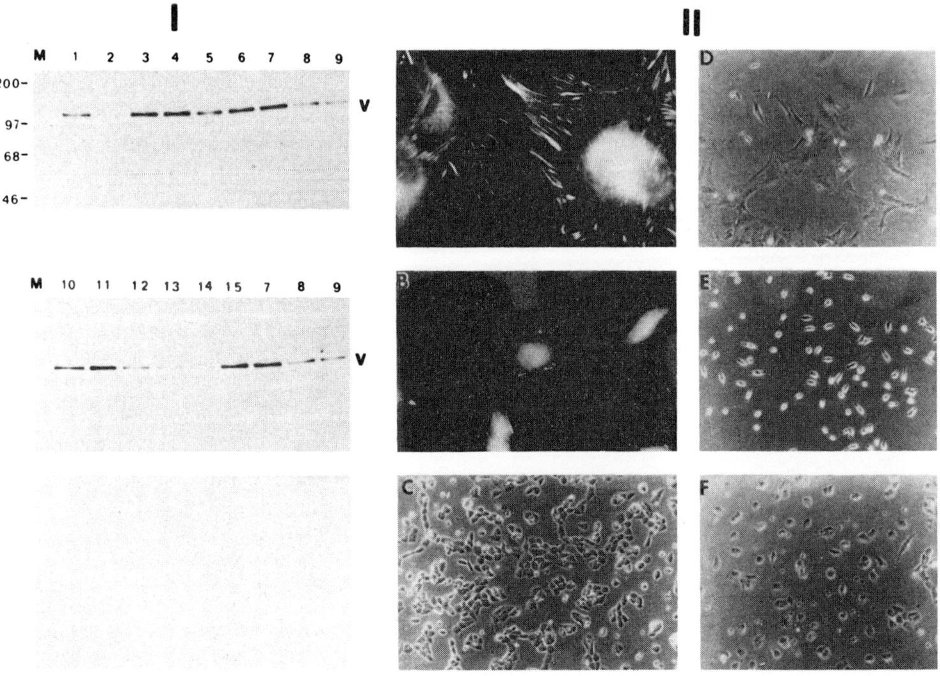

Figure 3. Antisense vinculin transfection confers a decrease in vinculin content and 3T3 cell size. 3T3 cells were co-transfected with an antisense mouse vinculin cDNA expression vector, and with the neo^r gene. G418 resistant colonies were screened for reduced vinculin expression (**I**) and organization (**II**) by immunoblotting and immunofluorescence with anti vinculin antibody. Lanes 1-5 and 10-15 are independent clones transfected with antisense vinculin; lane 7, 3T3 control; lane 8, 3T3 one third the amount of protein loaded; lane 9, SV40-transformed 3T3 (SVT2) cells. v, vinculin (From Rodríguez Fernández et al., 1993).

150

vinculin positive plaques (Figure 3B compare to Figure 3A). The motile and anchorage dependent properties of cells expressing diminished vinculin levels were also affected (Table 1). Cells expressing vinculin at a level lower than 30% of control had a round phenotype and an increased motility, while cells expressing vinculin at 50% of control and above, had a flat morphology and motile properties indistinguishable from 3T3 cells (Table 1). Moreover, clones expressing diminished levels of vinculin were capable of anchorage independent growth (Table 1), and formed large colonies in agarose culture. The results suggest that suppression of vinculin expression confers changes in cell shape, motility and anchorage dependence of 3T3 cells which are similar to those observed in transformed 3T3 fibroblasts (Ben-Ze'ev, 1985).

Table 1. Morphology, motility, and growth properties of antisense-vinculin transfected clones

Cell line	Vinculin levels % of 3T3[1]	Morphology on plastic	Colonies in agar/10^4 cells[2]	Migration cells/mm[3]
3T3	100	flat	30	180±20
D31	108.3	flat	ND	ND
A41	106.7	flat	4	210±60
C11	102.7	flat	ND	150±60
C41	98.8	flat	20	ND
D51	78.9	flat	ND	ND
A11	69.6	flat	ND	ND
D41	58.4	flat	6	ND
C51	43.4	flat	ND	ND
A51	31.6	round	188	400±80
B51	16.6	round	575	460±100
B61	11.0	round	143	360±50
C61	10.3	round	113	ND

[1] The levels of vinculin were determined by quantitative computerized scanning of the immunoblots shown in Fig. 3. The level of vinculin in 3T3 cells was taken as 100%.
[2] The number of colonies containing >50 cells was determined 2 weeks after seeding 10^4 cells (in duplicates) in agarose-containing medium.
[3] Cell motility was determined by counting the number of cells which migrated into an area of 1 mm^2 16 hours after introducing a "wound" into a confluent monolayer (From Rodríguez Fernández et al., 1993).

RESTORATION OF VINCULIN EXPRESSION IN MALIGNANT CELLS SUPPRESSES THEIR TUMORIGENICITY

Tumor cells in culture are characterized by altered growth, adhesion, shape and motility (Ben-Ze'ev, 1985). Among the more conspicuous morphological changes observed in transformed cells are the round shape and the diminished number of microfilaments and adhesion plaques (Pollack et al., 1975; Ben-Ze'ev, 1985, Raz and Ben-Ze'ev, 1987). The decrease in the number of microfilament bundles of transformed cells is often accompanied by a lower level of microfilament-associated proteins including, vinculin (Raz et al., 1986; Rodríguez Fernández et al., 1992b), α-actinin (Glück et al., 1993), gelsolin (Vandekerkhove et al., 1989), and tropomyosin (Matsumura and Yamashiro Matsumura 1986). To address the cause and effect relationship between the alterations in cell growth rate and cell structure in transformed cells, we have selectively modulated the expression of AJ proteins in transformed cells and studied their effect on the transformed phenotype.

We have chosen to study SV40-transformed 3T3 cells (SVT2) which express over fourfold less vinculin (Rodríguez Fernández et al., 1992b) and the highly malignant and metastatic BSp73 ASML pancreatic adenocarcinoma (ASML) which does not contain detectable levels of vinculin RNA and protein (Raz et al., 1986; Rodríguez Fernández et al., 1992b). Clones expressing stably different levels of the transfected vinculin were isolated from both cell lines. The transfected vinculin in SVT2 cells co-localized with the endogenous protein in adhesion plaques, and in ASML cells the transfected vinculin was organized in small patches at the ventral cell part. SVT2 cells expressing the highest level of vinculin displayed a flatter phenotype. By electron microscopic examination, the contact sites with the substrate in transfected ASML cells were over 2 fold larger when compared to vinculin negative ASML cells (Rodríguez Fernández et al., 1992b).

Table 2.   Tumorigenicity of SVT2 clones expressing different vinculin levels

Clone	Vinculin Level[1]		Tumor Incidence (%)	Growth in Agar[2] (%)
	Chicken	Mouse		
3T3	0	750	0/6 (0)	< 0.01
neo 1	0	ND	6/6 (100)	11.0±0.8
neo 3	0	140	6/6 (100)	13.3±1.0
SVT2	0	135	6/6 (100)	12.3±0.5
D 34	344	170	1/6 (16)	5.7±1.0
D 43	135	ND	5/6 (83)	ND
D 41	80	ND	5/6 (83)	ND
D 44	670	160	0/6 (0)	3.6±1.5

Balb/c mice were injected with 5×10^6 cells per animal.
[1] Arbitrary O.D. units of the chicken vinculin levels were obtained by densitometer scanning of cell lysates separated by 2-D gel electrophoresis.
[2] Cells were seeded in duplicates at 250 and 10^3 cells per 35 mm dish in soft agar and the number of colonies (per cent) formed per plate was determined. ND, not done. (From Rodríguez Fernández et al., 1992b).

Different SVT2 clones expressing increasing levels of vinculin showed a progressively stronger suppression of the tumorigenic ability of the cells when injected into syngeneic animals (Table 2) and nude mice (Rodríguez Fernández et al., 1992b). SVT2 cells in which vinculin levels were close to that of nontumorigenic 3T3 cells showed a complete suppression of their tumorigenic ability in syngeneic animals (Table 2, clone D44). The growth of these cells in soft agar was also reduced. Transfected ASML clones expressing increasing levels of vinculin showed similar results. In clones expressing the highest levels of vinculin the metastatic ability to the lung of cells injected into the foot pad was completely suppressed, and the formation of the primary tumor was inhibited several fold. In clones expressing moderate and low levels of vinculin there was an inhibitory effect on the metastatic potential of the cells, but there was no apparent effect on the formation of the primary tumor (Rodríguez Fernández et al., 1992b). The inhibiton of the metastatic potential correlated directly with the level of vinculin in these cells.

The results thus demonstrated that modulations in the expression of vinculin can effectively suppress the tumorigenic and malignant metastatic ability of cells.

SUPPRESSION OF TUMORIGENICITY IN SVT2 CELLS AFTER TRANSFECTION WITH α-ACTININ

α-Actinin is an abundant actin-binding protein that crosslinks microfilaments in vitro and, like vinculin, is also a major component of AJ (Ginsberg and Geiger, 1991; Burridge et al., 1988). While the role of vinculin in mediating actin attachment to the membrane is believed to be indirect (via talin), it has recently been shown that α-actinin can bind to the cytoplasmic domain of the transmembrane β_1-integrin receptor (Ottey et al., 1990). This suggests that α-actinin may directly bind actin filaments to transmembrane integrin receptors. Previous studies have shown that α-actinin expression is modulated during growth activation in 3T3 cells and in regenerating liver (Glück et al., 1992), and during differentiation of different cell types (Rodríguez Fernández and Ben-Ze'ev, 1989; Ben-Ze'ev and Amsterdam, 1987). Furthermore, the expression of α-actinin is reduced over 6 fold in SV40-transformed 3T3 cells (SVT2) (Glück et al., 1993).

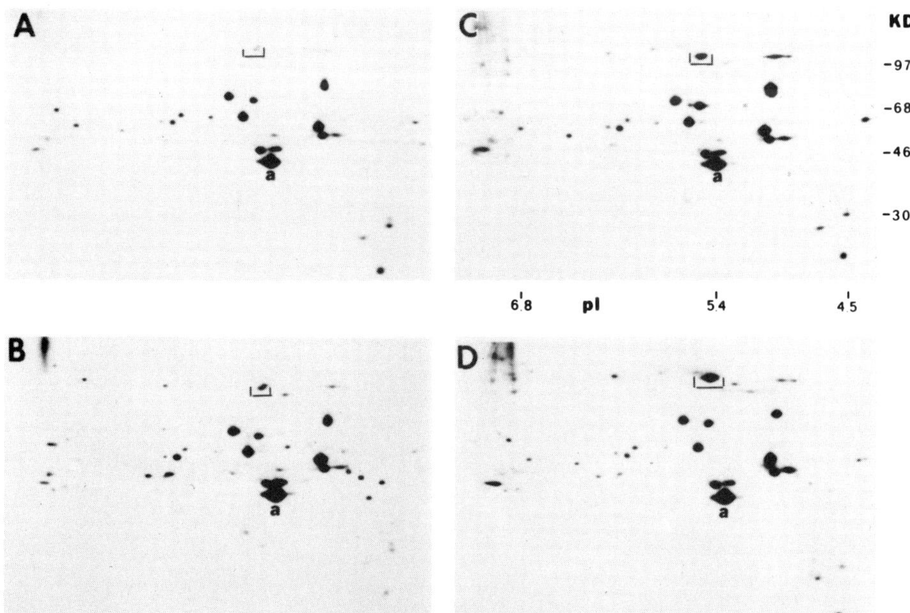

Fig. 4. Levels of α-actinin expressed by different SVT2 clones transfected with human α-actinin cDNA. SVT2 cells were co-transfected with the neo[r] gene and with a full length human α-actinin cDNA. Neo[r] control and clones expressing high levels of α-actinin were labeled with ^{35}S-methionine and analyzed by 2-D gel electrophoresis followed by autoradiography. a, actin. The bracket marks the position of α-actinin A, SVT2 neo[r] control; B-D, α-actinin transfected clones: B, Sα1; C, Sα8; D, Sα29.

To study the possible role of changes in α-actinin levels in cell transformation, we transfected SVT2 cells with a full length human cytoskeletal α-actinin which is 99% homologous to the mouse protein, and isolated clones expressing different levels of the transgene (Figure 4). Clones of SVT2 expressing between 4 (Figure 4B) to 10 fold (Figure 4D) higher levels of α-actinin were injected into syngeneic animals and nude, mice and the tumorigenic ability of the cells was determined (Table 3). The results show that as with vinculin, increasing α-actinin levels have a dramatic suppressive effect on the tumorigenicity of SVT2 cells. In cells expressing the highest level of α-actinin (about 2 fold higher than 3T3 cells) the tumorigenic ability was completely suppressed. Similar results were obtained with these clones injected into nude mice although tumor formation was not completely suppressed (Table 3). As with vinculin transfection, in SVT2 clones overexpressing α-actinin there was a significant decrease in the anchorage independence of these clones.

Table 3. Tumorigenicity of SVT2 clones expressing different α-actinin levels

Cell line	α-Actinin level[1]	Tumor incidence in Balb/C mice	(%)	Growth in agar (%)[2]	Tumor size in nude mice (cm^3)[3]
3T3	658.0	0/6	(0)	<0.01	0
SVT2	108.8	6/6	(100)	12.3±0.9	ND
Sneo	135.9	4/4	(100)	13.3±1.0	0.47±0.13
Sa16	250.1	4/5	(80)	11.5±0.8	ND
Sa1	497.3	2/4	(50)	ND	0.15±0.02
Sa8	693.0	1/5	(20)	5.1±0.7	ND
Sa29	1297.6	0/5	(0)	1.8±0.1	0.11±0.01

[1] Arbitrary O.D. units obtained by computerized densitometer scanning of 2-D gels as shown in Fig. 4.
[2] % number of colonies formed in agar compared to the number of colonies formed on plastic per same number of cells seeded.
[3] The size of tumors was determined 12 days after subcutaneous injection of 5x10^5 cells. N.D., not done.

CONCLUSIONS

The studies presented here have demonstrated that the modulation in the level of cytoskeletal plaque proteins observed during different cellular processes can be physiologically relevant. Thus, modest increases in AJ protein expression can dramatically inhibit motility, and the restoration, in tumor cells, of vinculin and α-actinin expression to near normal levels, can effectively suppress the tumorigenic ability of cells. We suggest that the signals generated via such modulations in AJ protein levels are mediated through the co-operative interactions of the various components in AJ (Figure 5). Biochemical and cell biological studies have previously shown that the assembly of AJ and the subsequent anchorage of microfilaments to the membrane is a co-operative process based on the multiple binding domains of AJ proteins such as vinculin (Kreis et al., 1984, and Figure 5). According to this model, a relatively modest increase in vinculin expression may lead to the recruitment and assembly in to adhesion plaques of a large number of AJ components, which will result in a more stable adhesion plaque, increased adhesion, and reduced motility. In contrast, a large reduction in the expression of AJ components such as vinculin and α-actinin, may be necessary to disassemble AJ integrity and affect microfilament-membrane binding, as was observed in tumor cells.

Future studies aiming to produce doubly targeted somatic cells with vinculin and α-actinin null mutations will help in characterizing the molecular basis of the role of AJ protein expression in cell function. In addition, embryonic stem (ES) cells having targeted vinculin and α-actinin genes will help understanding the role of regulating AJ protein levels in vivo in animal development.

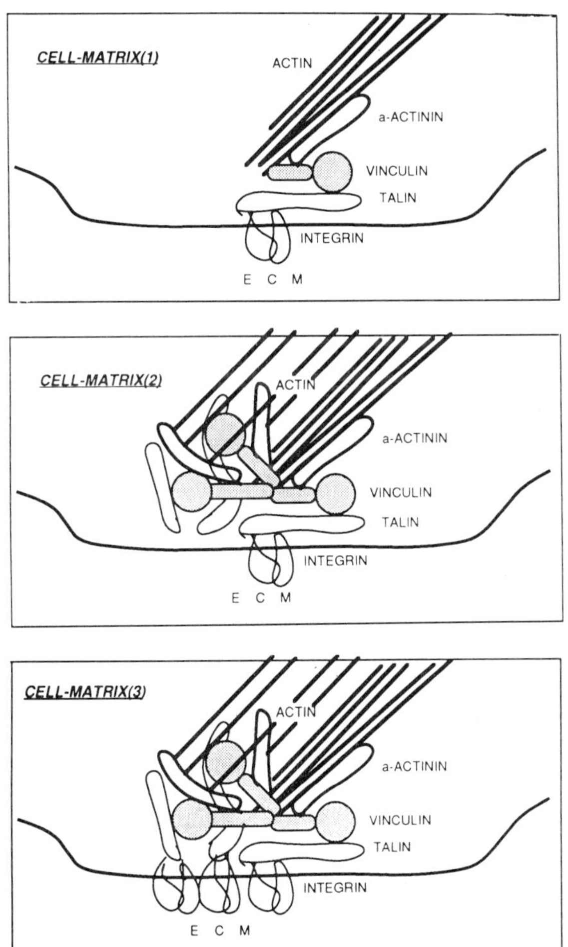

Fig. 5. Schematic models describing the co-operative molecular assembly of adhesion plaques. Cell-Matrix (1) shows a linear array of molecular interactions. Cell-Matrix (2) shows the cooperative effect resulting from vinculin oligomerization. Cell-Matrix (3) shows inside-out immobilization of transmembrane receptors to the newly formed contact sites.

ACKNOWLEDGMENTS

The studies described here were supported by grants from the Ministry of Science and Technology, Israel and the German Cancer Research Center, Heidelberg, Germany, from The Council For Tobacco Research-USA, from the Minerva Fund, from the Israel Cancer Research Fund (ICRF), and by a grant from the Leo and Julia Forchheimer Center for Molecular Genetics at The Weizmann Institute of Science. A. B-Z is the Lunenfeld-Kunin Professor for Genetics and Cell Biology. B. G is the E. Neter Professor for Cell and Tumor Biology.

REFERENCES

Belkin, A.M., O.I. Ornatsky, A.E. Kabanov, M.A. Glukhova, and V.E. Koteliansky. (1988). Diversity of vinculin/meta-vinculin in human tissues and cultivated cells. *J. Biol. Chem.* 263:14631-14635.

Bellas, R. E., R. Bendori, and S. R. Farmer. (1991). Epidermal growth factor activation of vinculin and β-integrin gene transcription in quiescent Swiss 3T3 cells. *J. Biol. Chem.* 266: 12008-12014.

Bendori, R., D. Salomon, and B. Geiger. (1987). Contact dependent regulation of vinculin expression in cultured fibroblasts: a study with vinculin specific cDNA probes. *EMBO J.* 6: 2897-2905.

Ben-Ze'ev, A. (1985). The cytoskeleton of cancer cells. *Biochim. Biophys. Acta.* 780: 197-212.

Ben-Ze'ev, A. (1991). Animal cell shape changes and gene expression. *BioEssays* 13: 207-212.

Ben-Ze'ev, A. (1992). Cytoarchitecture and signal transduction. *Crit. Rev. Eukaryotic Gene Exp.* 2: 265-281.

Ben-Ze'ev, A., and A. Amsterdam. (1987). In vitro regulation of granulosa cell differentiation: involvement of cytoskeletal protein expression. *J. Biol. Chem.* 262: 5366-5376.

Ben-Ze'ev, A., R. Reiss, R. Bendori, and B. Gorodecki. (1990). Transient induction of vinculin gene expression in 3T3 fibroblasts stimulated by serum growth factors. *Cell Regul.* 1: 621-636.

Burridge, K., K. Fath, T. Kelly, G. Nuckolls, and C. Turner. (1988). Focal Adhesions: Transmembrane junctions between the extracellular matrix and the cytoskeleton. *Ann. Rev. Cell. Biol.* 4: 487-525.

Burridge, K., C. E. Turner, and L. H. Romer. (1992). Tyrosine phosphorylation of paxillin and pp125[FAK] accompanies cell adhesion to extracellular matrix: a role in cytoskeletal assembly. *J. Cell Biol.* 119: 893-903.

Edelman, G. M. (1992). Mediation and inhibition of cell adhesion by morphoregulatory molecules. *Cold Spring Harb. Symp. Quant. Biol.* LVII 317-325.

Geiger, B., and D. Ginsberg. 1991. The cytoplasmic domain of adherens-type junctions. *Cell Motil. Cytoskel.* 20: 1-6.

Geiger, B., O. Ayalon, D. Ginsberg, T. Volberg, J. L. Rodríguez Fernández, Y. Yarden, and A. Ben-Ze'ev. (1992). Cytoplasmic control of cell-adhesion. *Cold Spring Harb. Symp. Quant. Biol.* LVII 631-642.

Glück, U., J. L. Rodríguez Fernández, R. Pankov, and A. Ben-Ze'ev. (1992). Regulation of adherens junction protein expression in growth-activated 3T3 cells and in regenerating liver. *Exp. Cell. Res.* 202: 477-486.

Glück, U., D. J. Kwiatkowski, and A. Ben-Ze'ev. 1993. Suppression of tumorigenicity in simian virus 40-transformed 3T3 cells transfected with α-actinin cDNA. *Proc. Natl. Acad. Sci. USA* 90: 383-387.

Guan, J-L., and D. Shalloway. (1992). Regulation of focal adhesion associated protein tyrosine kinase by both cellular adhesion and oncogenic transformation. *Nature* 358: 690-692.

Hynes, R.O. (1992). Integrins: versatility, modulation, and signaling in cell adhesion. *Cell* 69:11-25.

Juliano, R. L., and S. Haskill. 1993. Signal transduction from the extracellular matrix. *J. Cell Biol.* 120:577-585.

Kreis, T. E., Z. Avnur, J. Schlessinger, and B. Geiger. (1984). Dynamic properties of cytoskeletal proteins in focal contacts. *In* Molecular Biology of the Cytoskeleton. G. Borisy, D. Cleveland, and D. Murphy, editors. Cold Spring Harbor Laboratory, Cold Spring Harbor, NY. 45-57.

Matsumura, F., and Yamashiro-Matsumura, S. (1986). Tropomyosin in cell transformation. *Cancer Rev.* 6:21-39.

Moiseyeva, E. P., P. A. Weller, N. I. Zhidkova, E. B. Corben, B. Patel, I. Jasinska, V. E. Koteliansky, and D. R. Critchley. (1993). Organization of the human gene encoding the cytoskeletal protein vinculin and the sequence of the vinculin promoter. *J. Biol. Chem.* 268:4318-4325.

Otey, C.A., F. M. Pavalko, and K. Burridge. (1990). An interaction between a-actinin and the b1 integrin subunit in vitro. *J. Cell Biol.* 111:721-729.

Pollack, R ., M. Osborn, and K. Weber. (1975). Patterns of organization of actin and myosin in normal and transformed cells. *Proc. Natl. Acad. Sci. USA* 72:994-998.

Raz, A., and A. Ben-Ze'ev. (1987). Cell contact and architecture of malignant cells and their relationship to metastasis. *Cancer Met. Rev.* 6:3-21.

Raz, A., M. Zöller, and A. Ben-Ze'ev. (1986). Cell configuration and adhesive properties of metastasizing and nonmetastasizing Bsp73 rat adenocarcinoma cells. *Exp. Cell Res.* 162: 127-141.

Rodríguez Fernández, J. L., and A. Ben-Ze'ev. (1989). Regulation of fibronectin, integrin and cytoskeletal expression in differentiating adipocytes: inhibition by extracellular matrix and polylysine. *Differentiation* 42: 65-74.

Rodríguez Fernández, J. L., B. Geiger, D, Salomon, I. Sabanay, M. Zöller, and A. Ben-Ze'ev. (1992a) Suppression of tumorigenicity in transformed cells after transfection with vinculin cDNA. *J. Cell Biol.* 119: 427-438.

Rodríguez Fernández, J. L., B. Geiger, D. Salomon, D. and A. Ben-Ze'ev. (1992b). Overexpression of vinculin suppresses cell motility in Balb/C 3T3 cells. *Cell Motil. Cytosk.* 22: 127-134.

Rodríguez Fernández, J. L., Geiger, B., Salomon, D., and Ben-Ze'ev, A. (1993). Suppression of vinculin expression by antisense transfection confers changes in cell morphology, motility, and anchorage dependent growth of 3T3 cells. *J. Cell Biol.* in press.

Schwartz, M. A. (1992). Transmembrane signaling by integrins. *Trends Cell Biol.* 2: 304-308.

Takeichi, M. (1991). Cadherin cell adhesion receptors as a morphogenetic regulator. *Science* 251:1451-1455.

Ungar, F., B. Geiger, and A. Ben-Ze'ev. (1986). Cell contact and shape dependent regulation of vinculin synthesis in cultured fibroblasts. *Nature* 319: 787-791.

Vandekerckhove, J., G. K. Bauw, G. Vancompernolle, B. Honore, and J. Celis (1990). Comparative two dimensional gel analysis and microsequencing identifies gelsolin as one of the most prominent downregulated markers of transformed human fibroblast and epthelial cells. *J. Cell Biol.* 11:95-102.

Volberg, T., Y. Zick, R. Dror, I, Sabanay, C. Gilon, A. Levitzki, and B. Geiger. (1992). The effect of tyrosine-specific protein phosphorylation on the assembly of adherens type junctions. *EMBO J.* 11: 1733-1742.

Zachary, I., and E. Rozengurt. (1992). Focal adhesion kinase (p125^{FAK}): a point of convergence in the action of neuropeptides, integrins, and oncogenes. *Cell* 71:891-894.

Zieske, J. D., G. Bukusoglu, and I. K. Gipson. (1989). Enhancement of vinculin synthesis by migrating stratified epithelium. *J. Cell Biol.* 109: 571-576.

INDUCTION OF COLLAGEN SYNTHESIS IN RESPONSE TO ADHESION AND TGFβ IS DEPENDENT ON THE ACTIN-CONTAINING CYTOSKELETON

Jyotsna Dhawan and Stephen R. Farmer

Department of Biochemistry
Boston University School of Medicine
Boston, MA 02118

INTRODUCTION

Adhesive interactions between cells and the extracellular matrix (ECM) are thought to influence gene expression. Previously, we showed that the regulation of $\alpha1$(I) collagen expression is anchorage-dependent in Swiss 3T3 fibroblasts (Dhawan and Farmer, 1990). Non-adhesive culture conditions established by suspending cells in methylcellulose containing media result in a suppression of procollagen synthesis. Replating these suspended cells onto a tissue culture dish surface rapidly activates procollagen synthesis, and serum factors are not required for this response. The changes in type I procollagen synthesis reflect the levels of $\alpha1$(I) collagen mRNA. Induction of procollagen synthesis during replating is the result of regulation at transcriptional and post-transcriptional sites (Dhawan et al., 1991). Collagen mRNAs are destabilized in suspended cells and restabilized when adhesive contacts are restored. Transcription of the $\alpha1$(I) collagen gene is suppressed in suspended cells and reactivated by 18 hours after reattachment. In addition, the 5' flanking region of the rat $\alpha1$(I) collagen gene was found to contain sequences that conferred adhesion-responsiveness to a CAT reporter gene.

Actin: Biophysics, Biochemistry, and Cell Biology
Edited by J.E. Estes and P.J. Higgins, Plenum Press, New York, 1994

These observations are consistent with at least two hypotheses. Specific integrin-mediated cell-ECM interactions might be responsible for the activation of collagen expression during replating. However, reattachment of suspended cells to surfaces coated with different ECM proteins was found to activate procollagen synthesis to similar extents (J. Dhawan, unpublished data). Thus, whatever signal is being generated by attachment is not specific to a particular ECM-ligand: integrin interaction. This observation suggests that alternatively, collagen expression might be regulated in response to the cytoskeleton-dependent changes in cell morphology that occur when suspended cells encounter a surface. Upon reattachment to a surface, collagen synthesis is rapidly activated and increases before any appreciable cell spreading. However, the rate of collagen synthesis continues to rise as cells continued to spread for up to 12 hours. At this time, when cells have adopted a well spread morphology, increased transcription of the collagen gene is detectable. Changes in cell shape are difficult to quantitate as they are accompanied by alterations in surface area, nuclear:cytoplasmic ratio, number of substrate attachment points and reorganization of the cytoskeleton. Drug-induced perturbations of the microfilament and microtubule networks can be used, however, to prevent cytoskeleton-dependent shape changes in reattaching cells (Ben-Ze'ev et al., 1980). Cytochalasin D treatment has been shown to down-regulate collagen synthesis in epithelial cells, suggesting a role for the integrity of the actin cytoskeleton in the control of collagen gene expression (Svoboda and Hay, 1987). Therefore, we tested the hypothesis that microfilament-dependent cell shape changes are important in the activation of collagen synthesis in reattaching cells.

Another question raised by our studies was whether collagen expression could in fact be stimulated in the absence of adhesive interactions. The response of adherent fibroblasts to transforming growth factor β - a potent activator of ECM genes including collagen type I - has been extensively studied. TGF-β is a member of a well-characterized family of growth and differentiation factors and was originally identified as a factor that allowed anchorage-independent proliferation of cells whose growth is normally anchorage-dependent. Subsequently, cell adhesion proteins and their receptors were shown to be among the cellular targets for the action of this factor (Ignotz and Massague, 1986; Ignotz et al., 1987; Ignotz and Massague, 1987). TGF-β responsive elements have been identified in the upstream regions of the α2(I)   (Rossi et al., 1988) and α1(I) (Ritzenthaler et al., 1991; Ritzenthaler et al., 1993) collagen genes. This factor stimulates a pleiotypic response which includes activation of proto-oncogenes and induction of growth in soft-agar. Clearly, TGFβ is able to overcome the anchorage-dependence of growth-associated gene expression. The question we asked was whether TGFβ could overcome the apparent adhesion-dependence of collagen expression, and if so is its action dependent on the microfilament system.

RESULTS

Cytochalasin D blocks the adhesion-induced rise in procollagen synthesis

Type I collagen is one of the most prominent adhesion-responsive proteins in mouse fibroblasts (Dhawan and Farmer, 1990). Synthesis of this protein is down-regulated during suspension of 3T3 cells in methylcellulose media for 72 h and it is reactivated when these non-adherent cells are attached to surfaces on which they can spread. As we showed previously, two-dimensional gel electrophoresis is a convenient method for the estimation of relative rates of collagen synthesis (Dhawan and Farmer, 1990). The spot representing Type I collagen labelled C in Figure 1 was identified by three criteria: 1) It is the major proline-labelled cell-associated and secreted protein; 2) both of these proteins are collagenase sensitive, generating two acidic peptides consistent with the non-triple helical pro-peptides; 3) immunoprecipitation of [^{35}S] methionine-labelled extracts with anti-type I collagen antibody yields species that migrate with the same electrophoretic mobility as the major [^{14}C] proline-labelled species (data not shown).

Adhesion of suspended cells to type I collagen and fibronectin matrices induces collagen expression to similar extents - a 20 fold induction is seen 18 hours after replating on these adhesive surfaces, but increased synthesis is detectable as early as two hours after attachment. Reattachment of suspended cells to these ECM substrates did not induce the level much above that measured on tissue culture plastic (data not shown). This observation suggests that, rather than ligand-specific effects of integrin-matrix interactions, the attachment of the cell to any adhesive surface results in the reactivation of collagen expression, perhaps by re-organization of the cytoskeleton by tethering of ECM receptors. Previous studies suggested that a key regulatory mechanism is dependent on cell spreading since adhesion to different ECM molecules had similar effects on collagen expression. However, it is not possible to completely rule out a role for particular cell-ECM interactions due to the existence of a cage of fibronectin surrounding suspended cells. This pericellular matrix could potentially obscure any differences, as it is likely to be involved in initial interactions during replating, regardless of the composition of the matrix film coated on to the tissue culture plate. To analyze the role of the actin-containing cytoskeleton in the activation of collagen expression that accompany reattachment and spreading, we used the microfilament disrupting drug, cytochalasin D. Suspended cells were replated on to tissue culture plastic in serum-free medium in the presence or absence of cytochalasin D for 6 hours. During the last hour of replating, cells were pulsed with [^{35}S] methionine and total cell-associated proteins were analyzed by two-dimensional gel electrophoresis, shown in Figure 1. Changes in procollagen synthesis were quantitated by scanning the fluorograms using the 2-D software on a Bioimage Visage 60 densitometer. As expected, replating of suspended cells for 6 hours activated a 10-fold increase in the rate of procollagen synthesis. However, this rise was inhibited when cytochalasin D was added to the replating medium. Cytochalasin D-treated cells attached as rapidly as untreated cells, but did not spread appreciably during the time analyzed. Replating in the presence of colchicine, a microtubule-disrupting drug, also resulted in an inhibition of spreading but had no significant effect on procollagen synthesis (data not shown). Thus, it is likely that the organization of the actin cytoskeleton is important for the activation of collagen expression in reattaching cells and not cell spreading per se. It should also be noted that addition of cytochalasin D during replating significantly induced the synthesis of actin, the major microfilament constituent, and vinculin, an actin-associated protein found predominantly in adhesion plaques (see Figure 1, V=vinculin and A= actin).

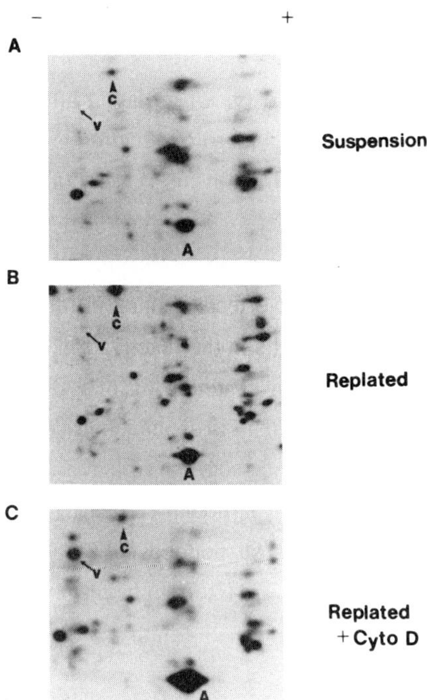

Figure 1. Cytochalasin D blocks the adhesion-induced rise in procollagen synthesis. Cells were suspended for 72 hours (A) in methylcellulose containing media with 10% serum and replated in serum-free medium in the absence (B) or presence (C) of cytochalasin D (5 ug/ml) for 6 hours. Proteins were labelled with [35S] methionine for 1 hour, isolated into lysis buffer A and displayed on 2-D gels as described in Dhawan and Farmer, 1990. Equal cpm (10^5) of the protein samples were analyzed. "C" marks the position of collagen, "v" indicates vinculin and "A" indicates actin.

TGF-β can Activate Collagen Expression in Suspended Cells

The observations documented above suggest that induction of collagen synthesis requires microfilament integrity. As suspended cells have low collagen synthetic rates, it was of interest to determine if collagen expression could be activated in these non-adherent cells by soluble effectors. TGF-β is known to activate ECM and integrin synthesis in

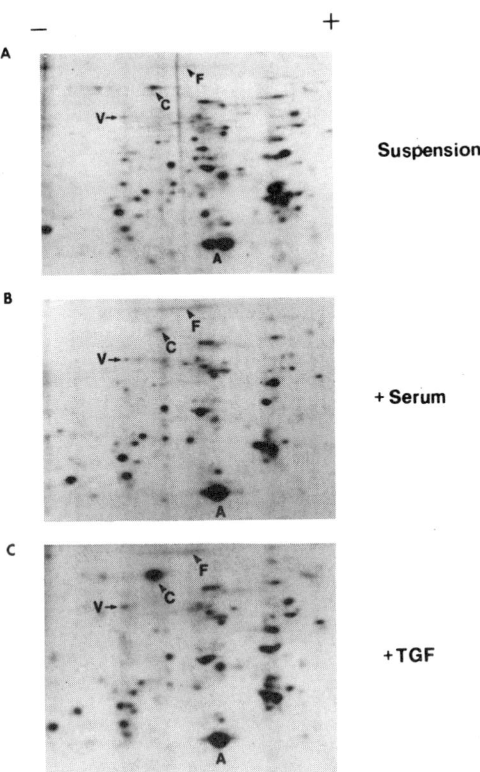

Figure 2. TGF-β activates procollagen synthesis in suspended cells. Suspended cells (A) were treated with 10% serum (B) or 3 ng/ml TGF-β1 (C) for 6 hours and newly synthesized proteins were analyzed as in Figure 1. "C" = collagen; "V" = vinculin ; "A" = actin and "F" = fibronectin.

adherent cells. The ED-50 for stimulation of anchorage-independent growth by TGF-β in Swiss 3T3 cells is 3 ng/ml (Ignotz and Massague, 1986). Suspended cells were therefore treated with 3 ng/ml TGF-β or 10 % calf serum for 6 hours and then pulse-labelled with ^{35}S-methionine. The radiolabelled proteins were analysed on 2-D gels. Figure 2 shows that TGF-β is very effective in inducing collagen synthesis in suspended cells (a > 10-fold induction was seen with 3 ng/ml in 6 hours of suspension).

Thus, TGF-β can overcome the anchorage dependence of collagen synthesis at the same dose that it activates anchorage-independent growth. As expected, serum did not induce collagen expression. In fact, procollagen synthesis was suppressed 3-fold in both suspended and adherent cells by serum addition.

Effect of Increasing Concentrations of TGF-β on Procollagen Synthesis

Suspended cells were treated with 0, 1, 3, 5 or 10 ng/ml of TGF-β for 24 hours. Cells were pulsed with 100 uCi/ml of [35S] methionine for the last hour and total cell-associated proteins were analyzed by 2D-PAGE as shown in Figure 2. The fluorograms were scanned and Figure 3 shows the dose-dependent rise in procollagen synthesis in response to TGF-β treatment. Three ng/ml is half as effective as 10 ng/ml in inducing collagen synthesis. This induction is a specific effect, as TGF-β does not markedly increase the total incorporation of [35S] methionine into proteins. Confluent cells were not as responsive to TGF-β as suspended cells (data not shown), suggesting that cell-cell contact might also affect the ability of cells to respond to this factor.

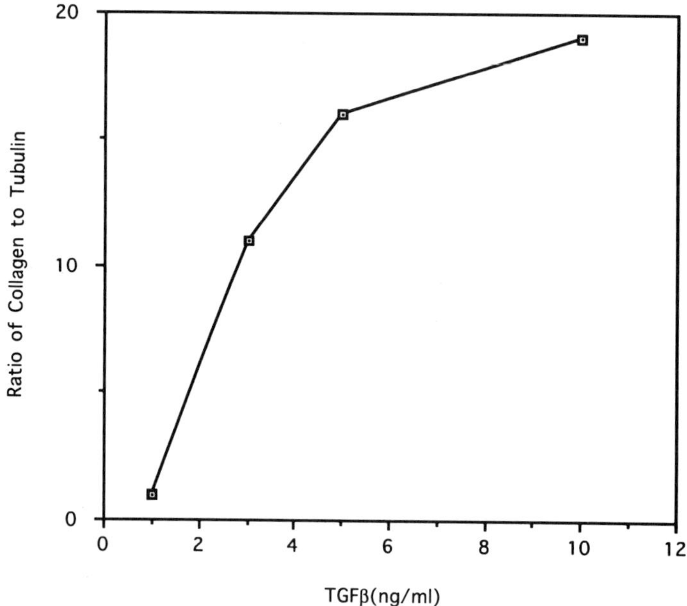

Figure 3. The effect of increasing concentrations of TGF-β on the expression of procollagen synthesis in suspended cells. Suspended cells were treated with TGF-β at a concentration of 1, 3, 5 or 10 ng/ml for 24 hours. Proteins were labelled with [35S] methionine and equal counts were analyzed by 2-D PAGE as shown in Figure 2. The fluorograms of each gel were scanned using a Bioimage Visage 60 densitometer and the integrated area under the peak representing collagen was normalized to the signal for β-tubulin.

Microfilament organization plays a role in the activation of collagen expression by TGF-β

The experiments described above demonstrate that replating of suspended cells results in high levels of collagen synthesis, perhaps due to microfilament reorganization triggered by adhesive interactions that are transduced across the membrane at sites of substratum contact. TGF-β is able to overcome the requirement for adhesive interactions as it can induce procollagen synthesis in suspended cells. One question raised by these studies is whether cytoskeletal organization is important for the response to TGF-β as well as to adhesion. TGF-β is known to increase the incorporation of ECM proteins into the matrix. It has long been known that a bidirectional integration of information exists at sites of ECM attachment to cells via integrins, i.e binding of fibronectin to the cell surface can induce the polymerization of the actin cytoskeleton. Therefore, it is possible that TGF-β action may be affected by microfilament organization. Suspended cells were therefore treated with TGF-β alone, cytochalasin D alone or TGF-β plus cytochalasin D for 6 hours and newly synthesized proteins analyzed by 2D-PAGE. Figure 4 shows that the ability of TGF-β to activate collagen expression in suspended cells is completely blocked by cytochalasin D. It appears therefore, that the regulation of collagen synthesis by this soluble factor is also dependent on an intact microfilament system.

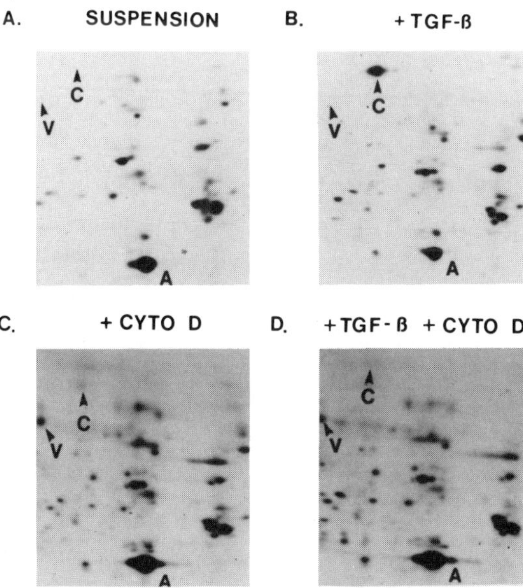

Figure 4. Cytochalasin D blocks the induction of procollagen synthesis by TGF-β in suspended cells. Suspended cells (A) were treated with 3 ng/ml TGF-β for 6 hours in the absence (B) or presence of cytochalasin D (D). Panel C shows the effect of treatment with cytochalasin D alone. Cells were analyzed as in Figure 1. "C" =collagen , "V" = vinculin and "A" = actin.

Interestingly, although TGF-β alone had no effect on the synthesis of actin and vinculin in suspended cells, the synthesis of both these proteins was induced whenever cytochalasin D was added, either in the presence or absence of the growth factor. As shown in Figure 1, replating in the presence of cytochalasin D also activated synthesis of actin and vinculin, concomitant with the inhibition of procollagen synthesis. Thus, re-organization of microfilaments is able to affect the synthesis of a subset of proteins that are components of the actin-containing cytoskeleton.

DISCUSSION

Studies documented in this report address the potential role of the actin cytoskeleton in the mechanisms that regulate collagen expression. Depolymerization of microfilaments in replated cells was effective in blocking the induction of collagen synthesis. This might be argued to be the result of preventing a potentially important shape change. However, treatment of reattaching cells with colchicine prevented spreading but did not affect collagen synthesis (data not shown). Therefore, a microfilament-dependent change in cell shape appears to be responsible for the activation of collagen synthesis during reattachment. Treatment of suspended cells with cytochalasin D had no effect on the basal level of collagen synthesis. However, in all cases cytochlasin D was able to activate synthesis of actin and vinculin. The activation of actin synthesis has been shown to occur during the cytochalasin D-induced depolymerization of microfilaments (Tannenbaum and Brett, 1985; Sympson and Geoghegan, 1990) in adherent cells.

Both vinculin and actin are immediate early growth-response genes that are transcriptionally induced following exposure of quiescent fibroblasts to growth factors (Bellas et al., 1991). Collagen expression on the other hand is commonly down-regulated by serum growth factors and appears to be reciprocally regulated to growth (Dhawan and Farmer, 1990). One mechanism by which microfilament disruption might affect gene expression is by activating signal transduction pathways that regulate the activity of a nuclear regulatory factor such as the serum response factor. Genes such as actin, vinculin, fibronectin and the c-fos proto-oncogene, that are rapidly activated by agents that induce the G0-G1 transition may be activated by cytochalasin D through this mechanism.

TGF-β has been shown to rapidly activate movement of integrins to the cell surface from intracellular pools (Ignotz and Massague, 1987). Cell surface expression of other ECM receptors/components may also be similarly affected. Thus, the ability of TGF-β to activate collagen expression in the absence of gross morphological alterations might be related to its ability to affect matrix composition and structure. This growth-factor induced alteration of the pericellular matrix in suspended cells may trigger changes in cytoskeletal organization that are required or responsible for the activation of collagen synthesis. The ability of cytochalasin D to block the TGF-β response is consistent with this hypothesis.

It has been suggested that the ability of TGF-β to inhibit the proliferation of adherent NRK cells is due to its induction of collagen expression (Nugent and Newman, 1989). We found that TGF-β was able to induce anchorage-independent expression of type I collagen at the same dose as it was able to activate anchorage-independent proliferation of Swiss 3T3 cells. Therefore, collagen may only be growth-restrictive in adherent cells. In addition, incorporation of collagen into a pericellular matrix is probably required for its effects on growth. Although suspended cells can be induced to synthesize collagen, we have no information on their ability to secrete and crosslink this matrix protein.

The microfilament system has been implicated in the regulation of gene expression including the localization of translationally active mRNA and control of mRNA turnover (Kislauskis and Singer, 1992). Procollagen synthesis is rapidly activated in replated cells prior to detectable increases in steady state mRNA (Dhawan and Farmer, 1990) and may be accomplished by the association of collagen mRNA with microfilaments. Interestingly, we have found that collagen mRNAs were much more stable in replated than in suspended cells (Dhawan et al, 1991). Thus, cell spreading and cytoskeletal organization may affect mRNA turnover. The effects of two activators of collagen expression (adhesion and TGF-β) are abrogated by cytochalasin D. Thus, it is possible that a primary mode of action for both soluble and matrix factors is their effect on the organization of the actin-containing cytoskeleton.

REFERENCES

Bellas, R.E., Bendori, R., and Farmer, S.R., 1991, Epidermal growth factor activation of vinculin and β1-integrin gene transcription in quiescent Swiss 3T3 cells, *J. Biol. Chem.* 266:12008.

Ben-Ze'ev, A., Farmer, S.R., and Penman, S., 1980, Protein synthesis requires cell-surface contact while nuclear events respond to cell shape in anchorage-dependent fibroblasts, *Cell* 21:365.

Dhawan, J., and Farmer, S.R., 1990, Regulation of α1(I)-collagen gene expression in response to cell adhesion in Swiss 3T3 fibroblasts, *J. Biol. Chem.* 265:9015.

Dhawan, J., Lichtler, A.C., Rowe, D.W., and Farmer, S.R., 1991, Cell adhesion regulates pro-α1(I) collagen mRNA stability and transcription in mouse fibroblasts, *J. Biol. Chem.* 266:8470.

Ignotz, R.A. and Massague, J., 1986, Transforming growth factor-β stimulates the expression of fibronectin and collagen and their incorporation into the extracellular matrix, *J. Biol. Chem.* 261:4337.

Ignotz, R.A., Endo, T., and Massague, J., 1987, Regulation of fibronectin and type I collagen mRNA levels by transforming growth factor-β. *J. Biol. Chem.* 262:6443.

Ignotz, R.A., and Massague, J., 1987, Cell adhesion protein receptors are targets for transforming growth factor-β, *Cell* 51:189.

Kislauskis, E.H., and Singer, R.H., 1992, Determinants of mRNA localization. *Current Opinions in Cell Biology* , 4:975.

Nugent, M.A. and Newman, M.J., 1989, Inhibition of normal rat kidney cell growth by transforming growth factor-β is mediated by collagen, *J. Biol. Chem.* 264:18060.

Ritzenthaler, J.D., Goldstein, R.H., Fine, A., Lichtler, A., Rowe, D.W., and Smith, B.D., 1991, Transforming growth factor-β activation elements in the distal promoter regions of the rat α1 type I collagen gene, *Biochem. J.* 280:157.

Ritzenthaler, J.D., Goldstein, R.H., Fine, A., and Smith, B.D., 1993, Regulation of the α1 (I) collagen promoter via a transforming growth factor-β activation element, *J. Biol. Chem.* 268:13625.

Rossi, P., Karsenty, G., Roberts, A.B., Roche, N.S., Sporn, M.B., and deCrombrugghe, B., 1988, A nuclear factor I binding site mediates the transcriptional activation of a type I collagen promoter by transforming growth factor-β, *Cell* 52:405.

Svoboda, K.K.H., and Hay, E.D., 1987, Embryonic corneal epithelial interaction with exogenous laminin and basal lamina is F-actin dependent, *Dev. Biol.* 123:455.

Sympson, G.J., and Geoghegan, T.E., 1990, Actin gene expression in murine erythroleukemia cells treated with cytochalasin D, *Exp. Cell Res.* 189:28.

Tannenbaum, J., and Brett, J.G., 1985, Evidence for regulation of actin synthesis in cytochalasin D treated HepG2 cells, *Exp. Cell Res.* 160:435.

GELSOLIN EXPRESSION IN NORMAL HUMAN KERATINOCYTES IS A FUNCTION OF INDUCED DIFFERENTIATION

Suzanne B. Schwartz[1], Paul J. Higgins[2], Ayyappan K. Rajasekaran[3], and Lisa Staiano-Coico[1]

[1]Department of Surgery
[3]Department of Cell Biology and Anatomy
Cornell University Medical College
New York, NY 10021

[2]Department of Microbiology , Immunology, and Molecular Genetics
Albany Medical College
Albany, NY 12208

INTRODUCTION

The epidermis is a self-renewing tissue comprised mainly of keratinocytes that exhibit different degrees of maturation depending upon their location (i.e. basal or suprabasal) within the tissue [1,2]. Under non-perturbed conditions, the turnover rate within the epidermis is relatively slow and the keratinocytes exist primarily in a non-migratory mode with cells exhibiting attachment to the basement membrane and to other surrounding cells via families of surface receptors known as integrins and cadherins[3,4]. Upon partial or full thickness injury, keratinocytes become *activated* and assume a migratory phenotype[5]. This activation process includes modulation of integrin expression, actin reorganization and changes within the complement of actin-associated proteins that enable the keratinocyte to migrate over and through the provisional wound matrix to re-establish a complete epithelial barrier.

Normal human keratinocytes (NHKs) respond to both differentiation-promoting [i.e. Ca^{2+} switch, retinoid depletion,sodium n-butyrate (NaB)] and differentiation-

Actin: Biophysics, Biochemistry, and Cell Biology
Edited by J.E. Estes and P.J. Higgins, Plenum Press, New York, 1994

inhibiting (fibronectin) signals *in vitro* [6,7,8,9,10] culminating in a pattern of growth generally analogous to wound healing regeneration within the skin[11]. Reorganization of the actin microfilament network is one of the specific hallmarks of NHK differentiation under these conditions[10,12,13,14].

Actin filament organization and assembly is regulated by various proteins serving different functions. A multitude of actin-binding proteins have been described that play roles in assembly, disassembly and gelation of actin monomers[15,16,17,18,19,20]. These proteins can affect three-dimensional filament architecture by organization into tight bundles, loose bundles and orthogonal nets (Fig. 1)[15-20].

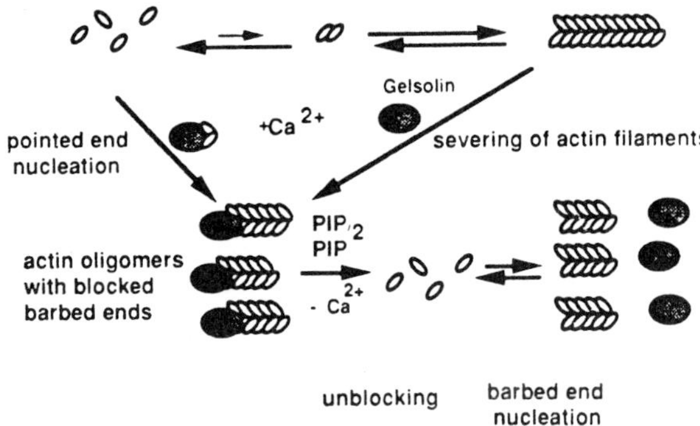

Fig. 1. Schematic model of actin filament assembly and its relationship to actin-associated proteins[15].(Reproduced with permission from T.P. Stossel, From Signal to Pseudopod, J Biol. Chem 1989)

Of particular importance is gelsolin, a 91 kDa calcium-dependent actin-associated protein that has several known functions including actin-severing, capping and nucleation activity [16]. Gelsolin has been implicated in cytoskeletal (CSK) rearrangements and changes in cell motility in a number of cell systems, including keratinocytes *in vitro* and *in vivo* [21,22,23,25]. Single cells suspended in methylcellulose undergo terminal differentiation within 24 hours[24]. Differentiation, in turn, is associated with a decrease in gelsolin expression. Differential expression of gelsolin is also observed within subpopulations of normal unwounded skin[25]. Peripheral membrane associated distribution of gelsolin is evident in the basal layer of the epidermis with more abundant and diffuse cytoplasmic expression observed in suprabasal keratinocytes[25]. Upon superficial wounding (i.e. blister formation), cellular gelsolin levels (assessed by immunofluorescence) appeared reduced in the migrating epithelial tongue, while a normal epidermal gelsolin distribution was retained in the unperturbed tissue surrounding the wound bed[25]. Qualitative changes in gelsolin expression have been described in stratifying colonies. Little is known, however, regarding quantitative changes in gelsolin expression during the growth and differentiation of NHKs in culture.

NHKs cultured under submerged conditions on plastic substrata (NHK/P) exhibit low levels of spontaneous cornified envelope (CE) formation[10,13,26]. Transition to mature CEs and the concomitant CSK rearrangements are augmented in NHK/P cultures during exposure to sodium n-butyrate (NaB). This system provides a model to probe underlying mechanisms behind the progressive CSK transitions observed in NHKs under differentiation induction conditions. The present studies were undertaken to quantitatively define the changes in CSK-actin and the actin-associated protein gelsolin in NHKs during growth before and after NaB-induced differentiation.

MATERIALS AND METHODS

Culture Of Human Epidermal Cells

Human epidermal cells were harvested from the trunks of cadaver donors, seeded onto either plastic or glass substrata, and grown in culture as previously described in MEM supplemented with penicillin/streptomycin (100 U/ml and 100mg/ml respectively), L-glutamine (1 mM), fungizone (0.25 mg/ml), 20% fetal calf serum (FCS) and 0.5 mg/ml hydrocortisone[10]. Cultures were grown at 37°C in a humidified 5% CO_2 atmosphere at an approximate density of 3×10^5 cells/cm^2. On days 2, 4, and 6 of growth, NHK cultures were fixed in situ using methanol-acetone (1:1) for analysis by fluorescence microscopy. On day 7 post-seeding a parallel series of cultures were treated with NaB (3mM). At daily intervals thereafter for 5 days, control and NaB-treated cultures were fixed as described above.

<u>Visualization of CSK-associated actin microfilaments (MF) and gelsolin distribution by fluorescence microscopy.</u> CSK actin organization was resolved in NHKs using rhodamine (Rh)-conjugated phalloidin as previously described[10]. Gelsolin expression was detected by indirect immunofluorescence utilizing an anti-human gelsolin antibody (1:100 dilution) (Ahmed and Higgins, in preparation). Prior to staining, cultures were post-fixed in 10% neutral-buffered formalin followed by incubation in 100% methanol at room temperature for 5 minutes. Cultures were washed three times in phosphate buffered saline (PBS), pre-incubated with PBS/2% BSA for 30 minutes at room temperature, incubated with gelsolin antibodies for 2 hr at room temperature in a humidified chamber, washed three times in PBS and incubated at room temperature with either TRITC-conjugated goat anti-mouse or FITC-conjugated goat anti-mouse (Sigma Chem Co., St. Louis, MO) secondary antibody

for 30 min. Cells were subsequently washed three times in PBS, and coverslips mounted in 50% glycerol/PBS for microscopic observation. Cells were examined using a confocal laser Nikon fluorescence microscope.Percent gelsolin-expressing NHKs was determined by manual quantitation of seven different fields per culture. Cells stained with FITC-conjugated secondary antibody were additionally double-labelled with propidium iodide (25mcg/ml;PI; Polysciences, Warrington, PA) to visualize nuclei for cytoarchitectural study via confocal laser microscopy.

<u>Flow Cytometric quantitation of CSK-actin and gelsolin in NHKs.</u> NHKs were harvested by trypsinization, cells were collected by centrifugation at 200 x g for 10 minutes, resuspended in PBS (at a concentration of 1 x 10^6 cells/ml) and fixed by vigorously pipetting 1 part cell suspension into 9 parts ice-cold methanol-acetone. Cells remained in fixative overnight prior to staining then centrifuged out of fixative at 200 x g for 10 minutes and washed three times in PBS. Cell suspensions were incubated in RNAse (5000U/ml; Sigma Chem. Co.) for 30 minutes at 37° C, centrifuged, and resuspended in either FITC-conjugated phalloidin (at 37°C), mouse anti-human gelsolin (1:300), or in normal mouse serum at (4°C) for 60 minutes with occasional agitation. After two washes in PBS, cells treated with FITC-phalloidin were counterstained with propidium iodide as described below. FITC- conjugated goat anti-mouse secondary antibody was added to the cell aliquots that had been treated with anti-gelsolin or normal mouse serum and incubated in the dark for 30 minutes at 4°C prior to counterstaining. After two washes in PBS, PI was added to the cells and incubated for 15-20 minutes on ice prior to measurement by flow cytometry.

RESULTS

Morphologic Transitions Of NHKs Before And After Differentiation In Response To NaB

Basal NHKs cultured under submerged conditions on plastic transit through defined morphologic states during growth in culture, resulting (by day 14) in the generation of stratified sheets of epidermal cells consisting of four to five cell layers[26,27] (Fig. 2) . The basal layer consists of rounded, but not tightly attached cells (Fig. 2A). The outer suprabasal layers is comprised of cells with increasingly condensed nuclei (Fig. 2B and 2C); 10± 4.6% of the upper layer surface in control cultures consisted of cornified envelope (CE)-like structures that were only infrequently enucleate. By contrast, NHKs treated for 4-7 days with 3mM sodium n-butyrate (NaB) took on a significantly altered appearance. Basal layer keratinocytes were highly basophilic and approximated only 50% of the size of control basal keratinocytes (Fig. 2D). A significant increase (58.6± 14.6%) was observed in the surface area occupied by CEs. Suprabasal maturation was a "mosaic" consisting of areas enriched in spinous cells (with particularly well-developed cytoplasmic filaments; Fig. 2E) and aggregates of nucleate pre-CEs and enucleate mature CEs (Fig. 2F).

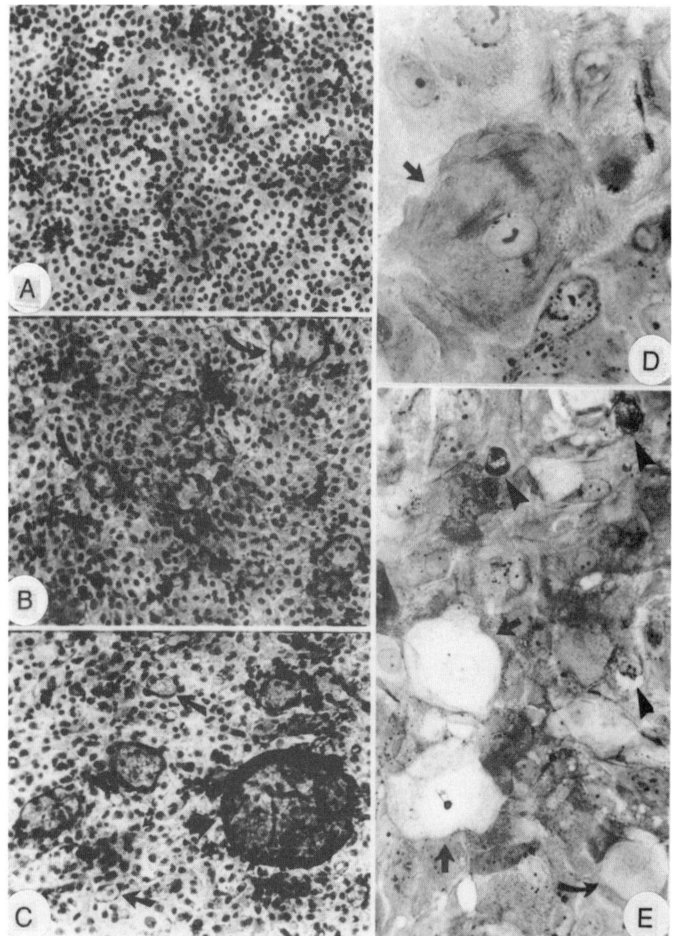

Fig. 2. Morphology of control and NaB-stimulated NHK populations as a function of culture substratum. (A) Typical morphology of NHK/P cultures approximately 1-day post confluency illustrating a

homogeneously-staining basal cell (i.e., high nuclear to cytoplasmic ratio; basophilic staining) population. At this early time point, there was no obvious distinction between NHKs cultured on uncoated tissue grade plastic or dried fibronectin (FN) matrices. (B) Cells grown to confluence on FN substrates then exposed to NaB (3mM) for 5 days exhibited some capacity for augmented differentiation compared to non-NaB-supplemented NHK/P or NHK/FN cultures as evidenced by an increase in the frequency of spinous NHKs (larger-than-basal cells with lower nuclear:cytoplasmic ratios and intracellular fibrous accumulations) and formation of immature CEs (curved arrows). In contrast, NHK/P-NaB populations (C) possessed not only a significant spinous compartment (36% of total culture area was occupied by spinous-like cells) but also abundant pre-CEs (straight arrows) as well as aggregates of fully mature CEs (large arrow). Regions of spontaneous epitheloid maturation were evident in NHK/P cultures but these involved, and were usually restricted to, spinous cell differentiation (arrow in D). By comparison, NHK/P-NaB cultures (E) exhibited the full range of differentiated phenotypes attainable in the submerged culture system. In one typical region of extensive epitheloid differentiation (E), several pre-CEs are evident (straight arrows; the bottom-most CE in this pair has a pyknotic nucleus) and one CE has extruded its nucleus (curved arrow); other nuclei in various stages of degeneration are also evident (arrowheads).

The morphologic changes that occur in NHKs during growth in the absence or presence of NaB are accompanied by dramatic alterations in the actin-based cytoskeleton. Two types of CSK-actin organization were consistently observed in control NHK cultures (Fig. 3). There were cells with dispersed, punctate aster-like actin-containing cables and cells with well developed peripheral band microfilaments (MF) with associated fine transcytoplasmic cables.

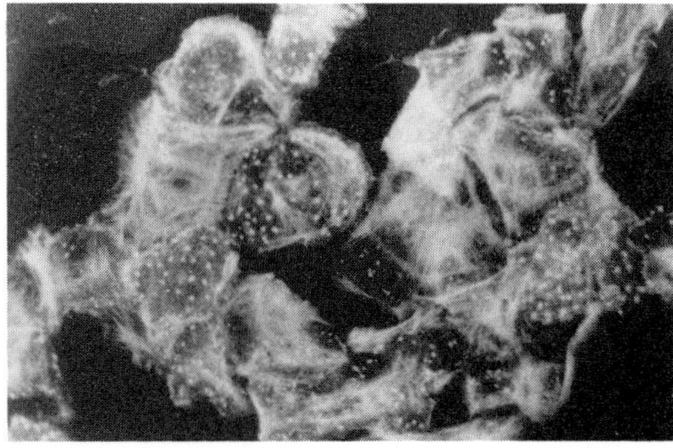

Fig. 3. CSK-actin distribution in control NHKs as visualized by rhodamine-phalloidin binding and u.v. microscopy.

Actin: Biophysics, Biochemistry, and Cell Biology
Edited by J.E. Estes and P.J. Higgins, Plenum Press, New York, 1994

revealed four putative CaM sites in the neck region of the heavy chain characterized by several hydrophobic and positively charged residues (Fig. 2). Other studies have localized some of the sites to the same region;[6,10-11] and the light chains of myosin IIs also bind in the neck region of myosin II heavy chains.[12] Mapping of the Ca^{2+}-stimulated phosphatidylserine-binding region was achieved with the use of a proteolytically modified BBMI heavy chain containing a 100-kDa fragment lacking the COOH-terminal 20 kDa of the heavy chain. As shown in Fig. 3A, BBMI with the truncated COOH-terminus retained Ca^{2+}-stimulated phosphatidylserine-binding similar to that observed for native BBMI. Ca^{2+}-independent phosphatidylserine-binding was reduced as expected, since the Ca^{2+}-independent phosphatidylserine-binding domain occurs in the COOH-terminal 20 kDa of the heavy chain.[6] Since the head domain of BBMI does not bind phosphatidylserine, the Ca^{2+}-stimulated phosphatidylserine-binding site must be localized within the neck region of the heavy chain amino-terminal to the Ca^{2+}-independent phosphatidylserine-binding region (Fig. 3B).

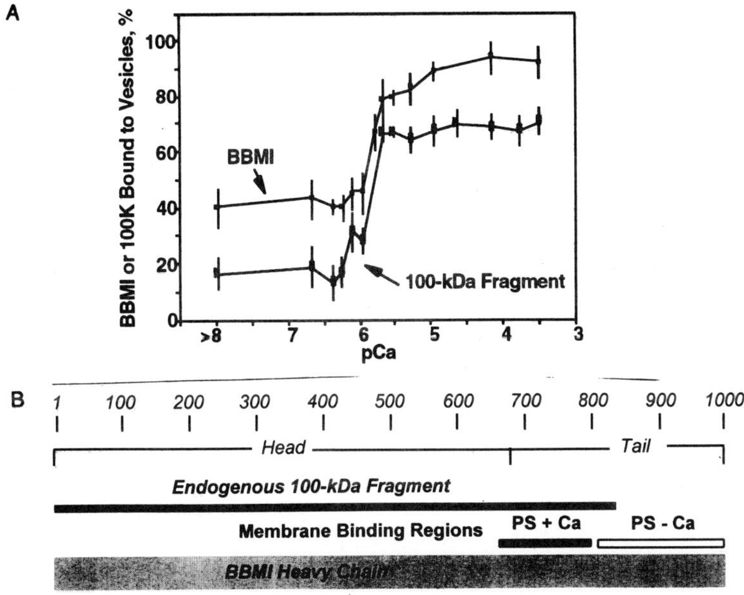

Fig. 3. Mapping the Ca^{2+}-induced phosphatidylserine binding site in BBMI. In *A*, mixtures containing a preparation of BBMI (0.43 mg/ml) in which 20% of the heavy chain was proteolyzed to an 100-kDa fragment were incubated with PKC (12 µg/ml) under standard conditions in the presence of PS and DOG at the indicated free Ca^{2+} concentrations in a volume of 35 µl for 2 min at 30 °C. The samples were centrifuged at 100,000 × *g* for 60 min and supernatant and pellet fractions were analyzed by SDS-PAGE. The distribution of the 100-kDa fragment of the BBMI heavy chain in the fractions was determined by densitometry of the gel and expressed as percent 100-kDa fragment bound to vesicles. Shown are averages ± S.D. for three experiments. In *B*, a schematic diagram of the BBMI heavy chain (*shaded rectangle*) is shown with its globular *head* and non-helical *tail* domains indicated. Aligned with the heavy chain is the 100-kDa fragment produced by endogenous protease (*Endogenous 100-kDa Fragment*). The Ca^{2+}-independent PS binding region (*PS - Ca*) is marked with an *open bar* and the Ca^{2+}-dependent PS binding site (*PS + Ca*) is marked with a *filled bar*. The amino acid positions along the heavy chain are numbered according to the sequence deduced from a partial cDNA clone (Garcia, A. et al. (1989) *J. Cell Biol.* **109**, 2895-2903) that most probably lacks about 40 NH_2-terminal residues, based on comparison with the complete bovine BBMI heavy chain sequence (Hoshimaru, M. et al. (1987) *J. Biol. Chem.* **262**,14625-14632). [Adapted with permission from *J. Biol. Chem.* **267**, 3445-3454 (1992)].

The CaM/PS Switch Is In a Distinctive Positively Charged Region

Examination of the distribution of charged amino acids in the heavy chain revealed that the neck region contains a stretch of almost exclusively positively charged residues with few negatively charged residues that is likely to participate in the binding of acidic CaM and phosphatidylserine (Fig. 4). Other myosins that contain CaM light chains, such as the product of the murine dilute locus,[13] have a homologous region of high net positive charge in the neck portions of their heavy chains (Fig. 5). A model of Ca^{2+}-stimulated membrane binding is shown in Fig. 6.

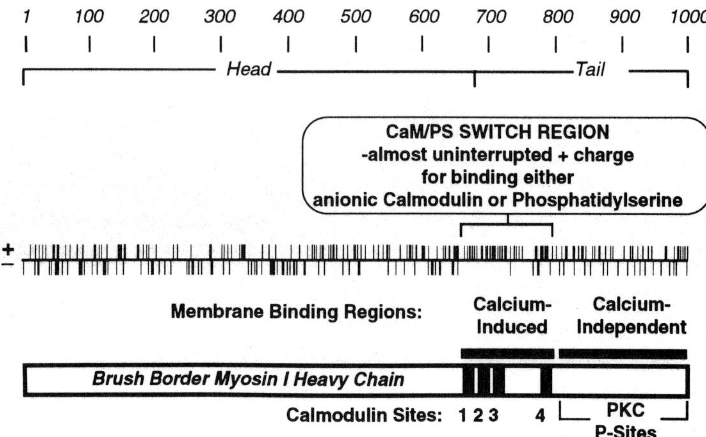

Figure 4. Identification of CaM/PS switch region within the BBMI heavy chain. A schematic diagram of the BBMI heavy chain (*open rectangle*) is shown. The *vertical bars in the rectangle* represent the four putative CaM binding sites (*CaM sites 1-4*). The Ca^{2+}-induced and Ca^{2+}-independent membrane binding regions are marked with *horizontal bars above the rectangle*. The region containing the PKC phosphorylation sites (*PKC P-sites*) is also indicated. Also shown is a line chart of the distribution of charged residues in the BBMI heavy chain, based on the sequence of BBMI from Garcia, A. et al. (1989) *J. Cell Biol.* **109**, 2895-2903. The position of the CaM/PS switch region is indicated. [Adapted with permission from *J. Biol. Chem.* **267**, 3445-3454 (1992)].

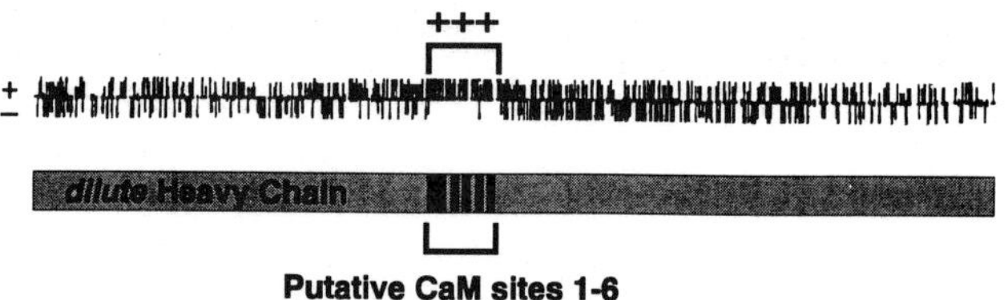

Figure 5. Identification of a putative CaM/PS switch region in murine *dilute* heavy chain. A schematic diagram of the murine *dilute* heavy chain (*shaded rectangle*) is shown. The *vertical bars in the rectangle* represent the six putative CaM binding sites. The position of the putative CaM/PS switch region is indicated above a line chart of the distribution of charged residues in the heavy chain based on the sequence of the protein encoded by the murine *dilute* locus from Mercer, J. A. et al. (1991) *Nature* **349**, 709-713. [Adapted with permission from *J. Biol. Chem.* **267**, 3445-3454 (1992)].

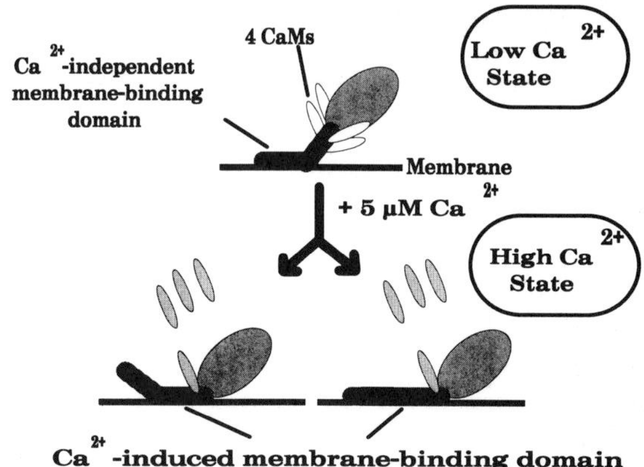

Figure 6. Model of the proposed CaM/PS switch of BBMI between two membrane-bound states. In the *Low Ca²⁺ State*, BBMI is bound to the membrane via the COOH-terminal half of its tail. BBMI is converted to its *High Ca²⁺ State* in the presence of micromolar Ca²⁺. In this process, Ca²⁺ first binds to CaM on the BBMI heavy chain, which induces the release of three CaMs from BBMI. The thus uncovered highly positively charged switch region of the heavy chain binds to the membrane. In this model two alternatives are shown, since it is not known whether the Ca²⁺-independent membrane binding domain remains associated with the membrane in the presence of micromolar Ca²⁺.

Ca²⁺-Regulation of BBMI ATPase Activity

The ATPase activity of BBMI in its soluble, non-membrane-associated state is regulated by Ca²⁺ in a highly temperature-sensitive manner.[1] At 30°C increasing Ca²⁺ from 10^{-7} to 10^{-5} M stimulates the actin-activated ATPase activity of BBMI by about 2.5-fold, as shown in Fig. 7. This stimulation is presumably triggered by the interaction of Ca²⁺ with the BBMI-bound CaMs, which all remain associated with the heavy chain under these conditions. At Ca²⁺ levels between 10^{-5}-10^{-4} M, one of the four CaMs dissociates from the heavy chain with no significant change in ATPase activity.

At 37°C the effects of increasing Ca²⁺ are strikingly different. Even at $< 10^{-7}$ M Ca²⁺, one CaM dissociates from BBMI (Fig. 8). An additional 1.5 CaMs dissociate at 10^{-6} M Ca²⁺ and no further dissociation occurs at up to 10^{-3} M Ca²⁺. The Ca²⁺-induced dissociation of CaM is accompanied by a loss of all actin-activated ATPase activity. Although not shown here, this is a consequence of an increase in the basal ATPase activity rather than of an inactivation of the enzyme.[1] These data are summarized schematically in Fig. 9 and a classification of CaM according to conditions needed for its dissociation from BBMI is presented in Table I. Loss of actin-activation with increasing Ca²⁺ and its reversal by exogenous CaM reported by others has led to the hypothesis that CaM dissociation might regulate BBMI ATPase.[2] However, it is clear from the retention of actin-activated ATPase activity and CaM at 30°C that Ca²⁺ can regulate BBMI ATPase without CaM dissociation. Studies of Ca²⁺ regulation of BBMI ATPase in membrane-associated states are in progress. Preliminary results show that ATPase activity is not affected by Ca²⁺-induced CaM dissociation in the presence of phosphatidylserine (Table I). Therefore, the loss of actin-activatable ATPase seen upon extensive CaM dissociation in the absence of phospholipid is apparently prevented by replacement of the CaM by phosphatidylserine.

A Second Actin-binding Site on BBMI is Indicated by BBMI-Actin Superprecipitation

Although *Acanthamoeba* myosins I have been shown to have an ATP-insensitive actin binding site in their tail domains,[14] BBMI has not been thought to have this type of site because

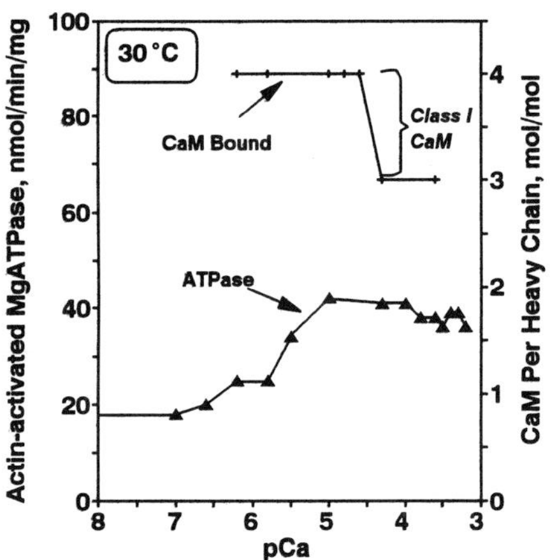

Figure 7. Effect of Ca^{2+} and myosin I CaM content on the Mg^{2+}-ATPase activity of BBMI at 30 °C. BBMI (35 µg/ml) was assayed at 30 °C for 60 min in solution containing 20 mM imidazole-HCl, 0.066 mM Tris-HCl, 1 mM ATP, 1.25 mM $MgCl_2$, 1 mM EGTA, 0-2.0 mM $CaCl_2$, pH 7.0, in the presence or absence of F-actin (0.36 mg/ml) in a volume of 125 µl. The actin-activated ATPase activity (*filled triangles*) was calculated by subtraction of the activity in the absence of actin from the activity in the presence of actin. The moles of CaM per mole of myosin I heavy chain (*pluses*) was determined by densitometry of SDS gels of 15-µl aliquots of supernates (*S*) and pellets (*P*) obtained after cosedimentation of myosin I (0.18 mg/ml) with actin (0.36 mg/ml) at 100,000 × *g* for 30 min at 30 °C. Prior to ultracentrifugation, myosin I and actin were incubated at 30 °C for 30 min in a solution containing 20 mM imidazole-HCL, 0.066 mM Tris-HCl, 1.25 mM $MgCl_2$, 1.0 mM EGTA, 0.045 M NaCl, and 0-2.0 mM $CaCl_2$ added to give the indicated free Ca^{2+} concentrations (*pCa*) at pH 7.0. The results presented are representative of six experiments obtained using three preparations of myosin I. [Adapted with permission from *J. Biol. Chem.* **266**, 1312-1319 (1991)]

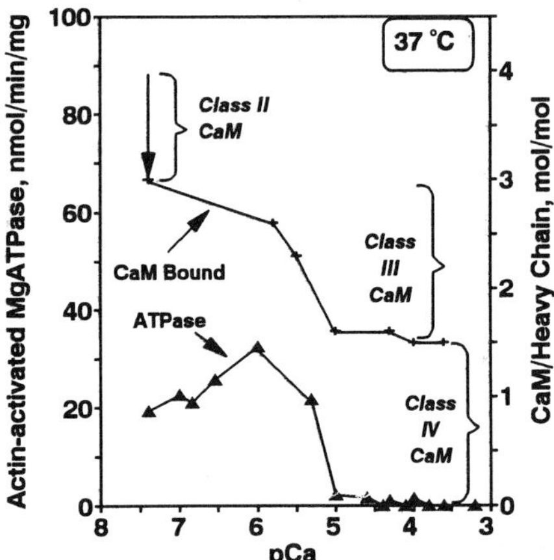

Figure 8. Effect of Ca^{2+} and myosin I CaM content on the Mg^{2+}-ATPase activity of BBMI at 37 °C. The experimental conditions are the same as for Fig. 7 except that the temperature was 37 °C. The loss of one CaM at 37 °C at pCa < 7 is indicated (*downward arrow*). The results presented are typical of those obtained in three separate experiments. [Adapted with permission from *J. Biol. Chem.* **266**, 1312-1319 (1991)].

the amino acid sequences of their tails are dissimilar and the kinetics of BBMI actin-activated ATPase is not triphasic. The ability of BBMI to superprecipitate actin[1] (Fig. 10), is apparently due to the presence of an ATP-insensitive actin-binding site in the tail (H. Swanljung-Collins, T. Ballard and J.H. Collins, unpublished). The stimulation of superprecipitation by 10^{-6} M Ca^{2+} parallels the Ca^{2+} stimulation of actin-activated ATPase seen at 30°C.

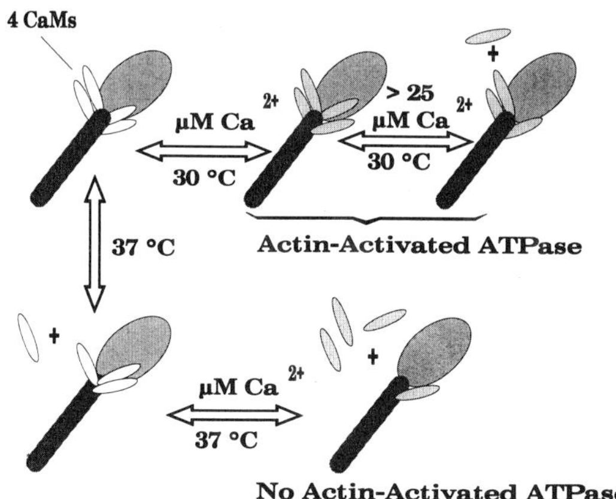

Figure 9. Summary of the effects of Ca^{2+} and temperature on the structure and ATPase activity of BBMI in its non-membrane bound state. The summary is of data in Figs. 7 and 8.

Table 1. Summary of the effects of Ca^{2+}, phospatidylserine and temperature on CaM dissociation from soluble and membrane-bound BBMI.

BBMI State	CaM	No.	Ca to Dissociate	Temp. to Dissociate	Effect of Dissociation on ATPase
Soluble	Class I	1	25-50 µM	30 °C	None
Soluble	Class II	1	None	37 °C	None
Soluble	Class III	1-2	1-10 µM	37 °C	Reversible Inhibition
Soluble	Class IV	1-2	?	?	?
Membrane Bound	PS-stimulated	3	1-5 µM	30 °C	None
Membrane Bound	PS-resistant	1	?	?	?

DISCUSSION

The multifunctional and multiligand-binding nature of the BBMI neck and tail is summarized schematically in Fig. 11. The Ca^{2+}-regulated CaM/PS switch described here for BBMI might be a common regulatory mechanism for all CaM-containing myosins, many of which are known to interact with membranes. It is noteworthy that the conformation of CaM bound to BBMI at low Ca^{2+} is likely to be different than the conformation of Ca^{2+}-saturated CaM present in Ca^{2+}-dependent target proteins. It will be interesting to find whether BBMI-bound CaM undergoes a conformational transition similar to that seen for CaM free in solution upon binding Ca^{2+}. While Ca^{2+} plays a major role in regulation of BBMI-membrane interaction and ATPase activity, the roles of phosphorylation of the tail by protein kinase C[4] and a BBMI tail kinase in regulation of these activities, as well as actin binding, remain to be established. It is significant that BBMI might be regulated by both Ca^{2+}/CaM and phosphoinositide signalling pathways. Finally, although the physiological role of tail actin binding is not yet clear, the presence of tail actin binding in a vertebrate myosin I like BBMI as well as in *Acanthamoeba* myosins I, suggests that tail actin-binding may serve an essential role in all myosins I.

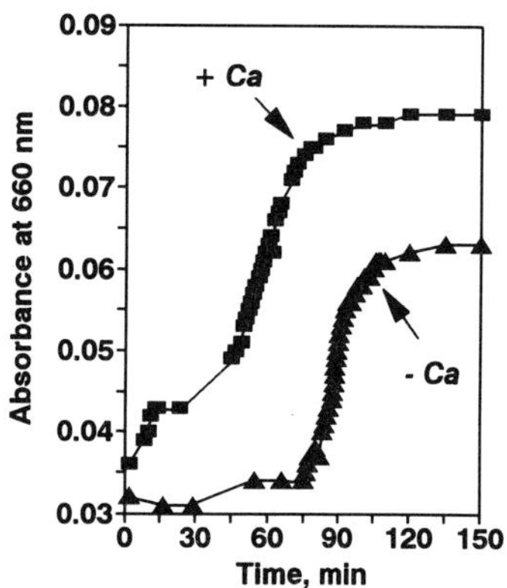

Figure 10. Ca^{2+} regulation of brush border actomyosin I superprecipitation. Chicken intestinal BBMI (0.18 mg/ml) and F-actin (0.36 mg/ml) were incubated at 22 °C in a total volume of 200 µl in a solution containing 20 mM imidazole-HCl, 0.066 mM Tris-HCl, 1 mM ATP, 1.25 mM MgCl$_2$, 1 mM EGTA, pH 7.0 and either 0.6 µM CaCl$_2$ (- *Ca^{2+}*) or 50 µM CaCl$_2$ (+ *Ca^{2+}*). Superprecipitation was measured by the absorbance at 660 nm. [Adapted with permission from *J. Biol. Chem.* **266**, 1312-1319 (1991)].

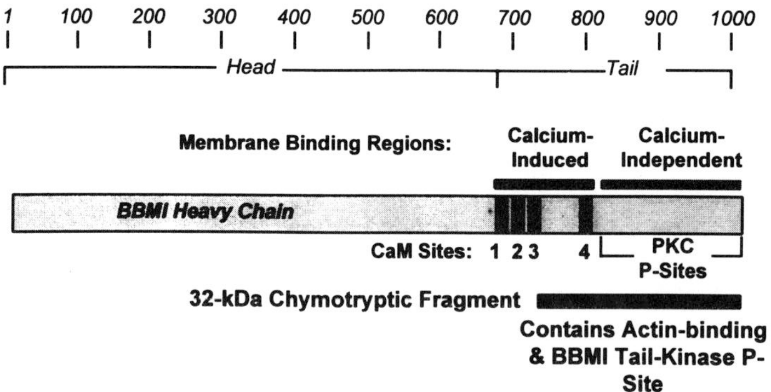

Figure 11. Schematic alignment of the actin-binding 32-kDa chymotryptic fragment with the heavy chain of chicken intestinal BBMI. This shows the relationship of the tail actin-binding region to the membrane binding regions, the CaM sites and the protein kinase C sites in the tail. The 32-kDa fragment also contains a site of phosphorylation by an intestinal brush border myosin I kinase.

ACKNOWLEDGMENTS

This work was supported by National Institutes of Health grant GM 32567.

REFERENCES

1. H. Swanljung-Collins and J.H. Collins, Ca^{2+} stimulates the Mg^{2+}-ATPase activity of brush border myosin I with three or four calmodulin light chains but inhibits with less than two bound., J. Biol. Chem. 266:1312 (1991).

2. K. Collins, J.R. Sellers, and P. Matsudaira, Calmodulin dissociation regulates brush border myosin-I (110K-calmodulin) activity *in vitro*, J.Cell Biol. 110:1137 (1990).

3. L.M. Coluccio and A. Bretscher, Calcium-regulated cooperative binding of the microvillar 110K-calmodulin complex to F-actin, J. Cell Biol. 105:325 (1987).

4. H. Swanljung-Collins and J.H. Collins, Phosphorylation of brush border myosin I by protein kinase C is regulated by Ca^{2+}-stimulated binding of myosin I to phosphatidylserine concerted with calmodulin dissociation, J. Biol. Chem. 267:3445 (1992).

5. K.A. Alexander, B. Cimler, K.E. Meier, and D.R. Storm, Regulation of calmodulin binding to P-57, J. Biol. Chem. 262:6108 (1987).

6. S.M. Hayden, J.S. Wolenski, and M.S. Mooseker, Binding of myosin I to phospholipid vesicles, J. Cell Biol. 111:443 (1990).

7. A. Garcia, E. Coudrier, J. Carboni, J. Anderson, J. Vandekerkhove, M. Mooseker, D. Louvard, and M. Arpin, Partial deduced sequence of the 110-kD-calmodulin complex of the avian intestinal microvillus shows that this mechanoenzyme is a member of the myosin-I family, J. Cell Biol. 109:2895 (1989).

8. E.R. Chapman, D. Au, K.A. Alexander, T.A. Nicolson, and D.R. Storm, Characterization of the calmodulin binding domain of neuromodulin: functional significance of serine 41 and phenylalanine 42, J. Biol. Chem. 266:207 (1991).

9. J. Baudier, J.C. Deloulme, A. Van Dorsselaer, D. Black, and H.W.D. Matthes, Purification and characterization of a brain-specific protein kinase C substrate, neurogranin (p17). Identificaiton of a consensus amino acid sequence between neurogranins and neuromodulin (GAP43) that correspond to the protein kinase C phosphorylation site and the calmodulin-binding domain, J. Biol. Chem. 266:229 (1991).

10. D.J. Halsall and J.A. Hammer, A second isoform of chicken brush border myosin I contains a 29-residue inserted sequence that binds calmodulin, FEBS Lett. 267:1:126 (1990).

11. M. Hoshimaru, Y. Fujio, K. Sobue, T. Sugimoto, and S. Nakanishi, Immunochemical evidence that myosin I heavy chain-like protein is identical to the 110-kilodalton brush-border protein, J. Biochem. 106:455 (1989).

12. L. Nyitray, E.B. Goodwin, and A.G. Szent-Gyorgy, Complete primary structure of a scallop striated muscle myosin heavy chain: sequence comparison with other heavy chains reveals regions that might be critical for regulation, J. Biol. Chem. 266:18469 (1991).

13. J.A. Mercer, P.K. Seperack, M.C. Strobel, N. Copeland, G., and N.A. Jenkins, Novel myosin heavy chain encoded by murine *dilute* coat colour locus, Nature 349:709 (1991).

14. E.D. Korn and J.A. Hammer 3d, Myosin I, Curr. Opinion Cell Biol. 2:57 (1990).

CONTROL OF p52(PAI-1) GENE EXPRESSION IN NORMAL AND TRANSFORMED RAT KIDNEY CELLS: RELATIONSHIP BETWEEN p52(PAI-1) INDUCTION AND ACTIN CYTOARCHITECTURE

Michael P. Ryan and Paul J. Higgins

Department of Microbiology, Immunology and Molecular Genetics
Albany Medical College
47 New Scotland Avenue
Albany, New York 12208

INTRODUCTION

Alterations in cell shape or substrate adhesion often accompany changes in the expression and/or distribution of proteins that influence cellular architecture; these include structural elements which comprise the extracellular matrix (ECM), their transmembrane receptors (integrins) as well as components of the focal adhesion sites (focal contacts) (Spiegelman and Farmer, 1982; Ben-Ze'ev, 1986; Ben-Ze'ev, 1987; Dike and Farmer, 1988; Rodriguez et al., 1989; Dalton et al., 1992). Modulation of cell morphology and adhesivity may be a direct consequence of specific perturbations within the actin cytoskeleton. The actin-based microfilament network undergoes dramatic reorganization after exposure of cells to transforming retroviruses (Altenburg et al., 1976; Wang and Goldberg, 1976), growth factors (Bockus and Stiles, 1984; Herman and Pledger, 1985; Ridley and Hall, 1992) and microfilament-disrupting agents such as the cytochalasins (Goodman and Miranda, 1978; Schliwa, 1982; Cooper, 1987). Such induced architectural changes frequently signal specific changes in gene expression and cell growth behavior. In normal rat kidney (NRK) cells, for example, actin reorganization (associated with transformation by retroviral oncogenes or cell shape-modulating drugs) is typically reflected in morphologic restructuring, in reduced substrate adhesion and in the reprogramming of gene expression (Ryan and Higgins, 1988, 1989, 1991; Higgins and Ryan, 1989a, 1989b; Higgins et al., 1991). One gene which appears particularly susceptible to shape-associated expression in the NRK cell system encodes the 52-kDa type-1 inhibitor of plasminogen activator [p52(PAI-1)].

After a brief introduction to the biology of the plasmin-based pericellular proteolytic system and its regulatory constituents (e.g., PAI-1), we will discuss actin cytoarchitecture with respect to control of p52(PAI-1) expression as well as the potential role of p52(PAI-

1) in regulation of cell shape. We will consider three distinct cell culture systems (i.e., flat revertants of transformed NRK cells and serum-stimulated or cytochalasin-treated NRK cells) which illustrate that modulation of the actin cytoskeleton in these cells is consistently associated with changes in p52(PAI-1) gene expression.

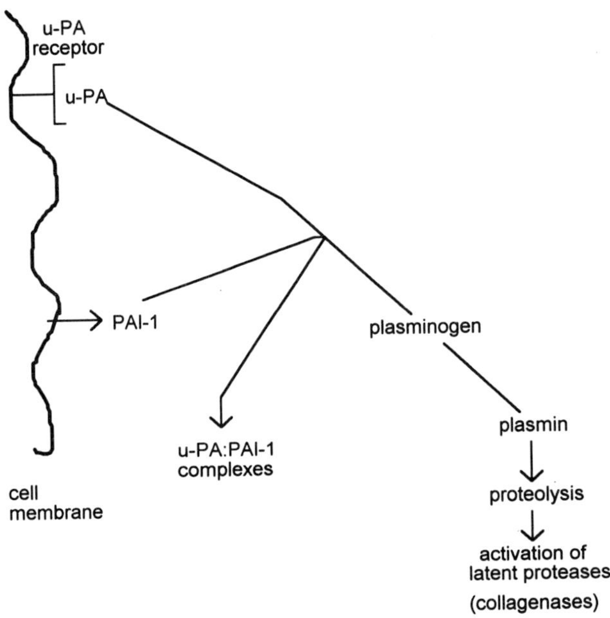

Figure 1. The plasminogen activation cascade. Conversion of the inactive zymogen plasminogen to the active broad spectrum protease plasmin is catalyzed by urokinase type plasminogen activator (u-PA) which is focalized at the cell surface by virtue of interaction with its specific receptor. Plasminogen activator inhibitor type-1 (PAI-1) specifically inhibits this reaction resulting in a net decrease in pericellular proteolytic activity (adapted from Laiho and Keski-Oja, 1989).

The Plasmin Proteolytic Cascade: Structure-Function Relationships and Directed Pericellular Proteolysis

Maintenance and turnover of the ECM through controlled pericellular proteolysis is a highly-regulated process involving various matrix-degrading proteases and their specific inhibitors. Within this content, the broad spectrum protease plasmin has been implicated in several physiological processes including thrombolysis, ovulation, cell migration and metastasis (Dano et al., 1989; Pollanen et al., 1991). In the plasminogen-dependent proteolytic cascade, urokinase (u-PA) and tissue-type (t-PA) plasminogen activators specifically convert the inactive zymogen plasminogen to plasmin; this activation results in generalized ECM proteolysis and activation of additional latent proteases (usually of the metallo-protease class). Plasminogen activator inhibitors (including the major physiological inhibitor of u-PA, i.e., PAI-1) are important negative regulators of this proteolytic cascade (Figure 1). Due to the localization of PAI-1 and u-PA to areas surrounding and within the region of the focal contact, respectively (Pollanen et al., 1987; Pollanen et al., 1991), these critical elements of the plasmin-based proteolytic cascade are ideally positioned to influence cell-to-substrate adhesion and, thereby, cell shape (Figure 2). This relationship is also valid for cells of the NRK lineage where p52(PAI-1) is deposited to the cellular ventral undersurface region localizing in a strial-like pattern consistent with its exclusion from the focal contact structure proper (Pollanen et al., 1987, Rheinwald et al., 1987). This is clearly evident upon co-visualization of p52(PAI-1) and actin using double label fluorescence microscopy since areas devoid of p52(PAI-1) immuno-staining co-localize to actin bundle termini (Figure 3). This ventral undersurface matrix "carpet" and associated structures are amenable to biochemical analysis. Focal contact-like structures can be isolated by treatment of substrate-adherent cells with saponin and subsequently detached by forced pipetting (Neyfakh and Svitkina, 1983). Attachment sites indicative of focal contacts (as confirmed by interference reflection microscopy) are retained after such treatment along with specific cellular membrane fragments. Proteins

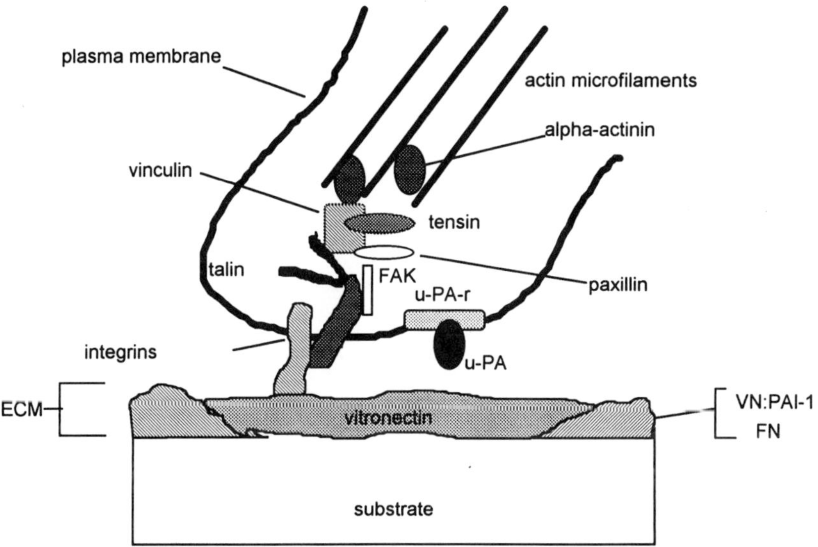

Figure 2. Idealized schematic illustrating localization of u-PA and PAI-1 at focal adhesion sites. The relationship between constituents of the focal contact and components of the plasmin-mediated proteolytic system demonstrate the potential importance of the plasmin-mediated proteolytic cascade as a regulator of cell-to-substrate adhesion. u-PA (associated with its receptor) and PAI-1 are localized to the extracellular face of the cell within and surrounding the focal contact, respectively (derived from Pollanen et al., 1991 and Ezzell, 1993).

typically localized to intact focal contact sites (i.e. vinculin), however, are not detected by immunofluoresence unless cells are removed by gentle rather than vigorous pipetting (Neyfakh and Svitkina, 1983). One-and two-dimensional electrophoretic analysis of the saponin-resistant (SAP) proteins of NRK cells revealed that p52(PAI-1) and actin are the major constituents of this fraction (Figure 4). Direct visualization of p52(PAI-1) by immunofluoresence in saponin-treated cultures confirm that p52(PAI-1) remains associated with the culture dish even after cell detachment emphasizing the potential of PAI-1, by virtue of its abundance and subcellular distribution, to function as a regulator of pericellular proteolysis in areas of cell-to-ECM attachment (Figure 4).

Sodium Butyrate-Induced p52(PAI-1) Expression and Cytoskeletal Reorganization in v-K-_ras_-Transformed NRK Cells

Oncogenic conversion of established fibroblasts by acute transforming retroviruses (or their isolated v-onc genes) is associated with extensive alterations in cell morphology and substrate adhesion (Matsumura et al., 1983; Wahrman et al., 1985; Ryan and Higgins, 1988). Cytostructural changes that typically accompany retroviral transformation of mammalian fibroblasts include the loss of stress fiber structure (Altenburg et al., 1976; Wang and Goldberg, 1976), altered expression of specific tropomyosin isoforms (Matsumura et al., 1983; Cooper et al., 1985), disruption of adhesion plaques (Maness,

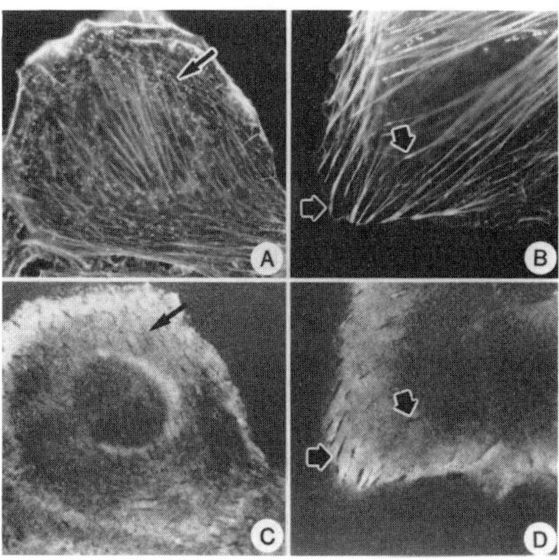

Figure 3. Double immunofluorescenct localization of actin and p52(PAI-1) in NRK cells. Rhodamine-conjugated phalloidin and anti-p52(PAI-1) antibodies were used to examine the distribution of actin (A and B) and p52(PAI-1) (C and D), respectively in formalin-fixed, NP-40 permeabilized NRK cells. NRK cells contain numerous transcytoplasmic actin filaments; actin bundle termini are clearly evident as thickened structures (arrows). p52(PAI-1) is present as a carpet beneath the cell with intense staining at the cell periphery. Areas negative for p52(PAI-1) staining co-localize with actin bundle termini (arrows) giving the strial-like pattern typical of PAI-1 ventral undersurface deposits.

1981), aggregation of the microfilament/adhesion plaque-associated proteins talin, vinculin, alpha-actinin and fimbrin (Carley et al., 1981; Stickel and Wang, 1987) and impaired cell-substratum contact (Altenburg et al., 1976; Hagman et al., 1981). NRK cells grown under normal culture conditions are epithelioid in morphology, contain numerous transcytoplasmic actin filaments, construct vinculin-containing adhesion plaques and are density-dependent for growth (Higgins and Ryan, 1989; Ryan and Higgins, 1993). Their Kirsten murine sarcoma virus-transformed derivatives (KNRK cells), in contrast, have a markedly abnormal cytoarchitecture and reduced substrate adhesion (Ryan and Higgins, 1988). Partial reversion to a more normal phenotype occurs during exposure of KNRK cells to the differentiating-inducing agent sodium butyrate (NaB). While KNRK cells are devoid of actin stress fibers and fail to form vinculin-containing adhesion plaques, NaB-treated KNRK (K/NaB) cells (continuous exposure to 2 mM NaB for 2-3 days) have a flattened morphology, display numerous fine transcytoplasmic actin filaments and exhibit vinculin-containing structures at the cell periphery which have interference reflection characteristics of focal contacts (Figures 5 and 6). Since these cyotarchitectural changes were accompanied by increases in cell-to-substrate adhesion, electrophoretic analysis of subcellular fractions enriched in substrate attached material (prepared by detachment of cells with EDTA) was utilized to identify specific proteins involved in NaB-induced phenotypic conversion (Figure 7). KNRK cultures were virtually devoid of p52(PAI-1); K/NaB cells accumulated substrate-associated p52(PAI-1) deposits at levels approaching that of NRK cells. 2-D electrophoretic separations of substrate-enriched protein fractions confirmed that the induced protein is identical in mobility and pI to p52(PAI-1) synthesized by NRK cultures (Higgins and Ryan, 1989a). Increased substrate deposition reflected an increase in steady state levels of p52(PAI-1) mRNA (Figure 8); induction, however, is inhibited by actinomycin D suggesting the requirement for ongoing RNA synthesis during the period of NaB exposure (Higgins and Ryan, 1991). Actin mRNA abundance, in contrast, did not change as a result of NaB exposure. Since KNRK cells fail to express both p52(PAI-1) mRNA and protein, it appears that the lack of detectable p52(PAI-1) deposition in cultures of v-<u>ras</u>-transformed cells is not simply a consequence of defective targeting (of the protein) into the substrate and that induction most likely involves reprogramming of p52(PAI-1) gene expression.

The KNRK cell system suggests that decreased p52(PAI-1) expression is linked to *ras*-mediated transformation and that reversal of the phenotype by NaB is mediated in part by p52(PAI-1). Similarly, NRK cells transfected with the activated EJras^{val-12} gene or possessing a temperature-sensitive lesion within the p21^{v-K-}*ras* display decreased p52(PAI-1) expression and aberrant morphology (Higgins et al., 1991). Reversal of the transformed phenotype in the temperature sensitive mutants by cultivation at the non-permissive temperature is associated with normal morphological characteristics and p52(PAI-1) content at levels comparable to that in NRK (Higgins et. al., 1991). That p52(PAI-1) down regulation is associated with ras-mediated transformation is clear; the contention that p52(PAI-1) is associated with cell shape regulation is supported by evidence that NRK cells undergo extensive cell flattening when plated onto SAP fraction matrices or upon addition of exogenously added SAP fraction proteins. SAP-induced spreading, however, is inhibited in the presence of anti-p52(PAI-1) antibodies (Higgins et al., 1991).

p52(PAI-1) Expression is Linked to NRK Cellular Growth State: Regulated Expression and Cytostructural Changes

Analysis of the K/NaB cell system implicates p52(PAI-1) induction and substrate deposition as an important aspect in either the formation or stability of focal contacts. A similar conclusion has been reached upon examination of specific PAI-1-binding components in human HT1080 sarcoma cells (Ciambrone and McKeown-Longo, 1990). It is likely that ECM-bound PAI-1 alters the proteolytic microenvironment at areas of cell-to-substrate adhesion. To examine this relationship within the context of cell growth activation, physiological conditions defined in which p52(PAI-1) could be induced in a regulated manner. Compared to quiescent NRK cultures, actively proliferating (subconfluent) and newly confluent cells are characterized by substantially greater levels of SAP-associated p52(PAI-1) protein suggesting that synthesis is regulated in a growth state-dependent manner (Ryan and Higgins, 1993). p52(PAI-1) expression is, in fact, associated with the growth state of NRK cells since p52(PAI-1) substrate deposition decreased under conditions of serum deprivation but is rapidly induced after addition of 20% FBS-containing media to serum-starved, quiescent cultures (Figure 9).

Activation of p52(PAI-1) expression as a function of serum stimulation occurs in an immediate early response manner since induction requires RNA synthesis but not _de novo_

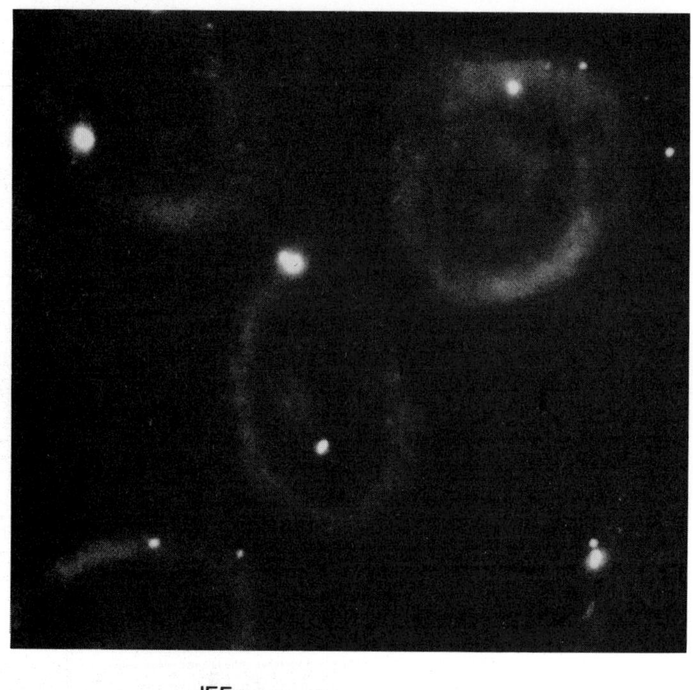

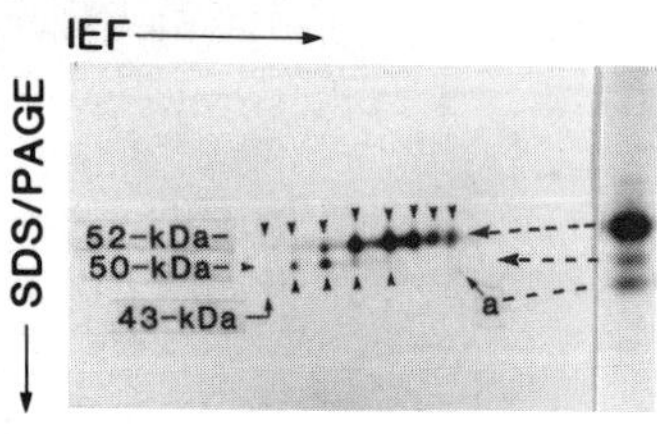

Figure 4. Association of p52(PAI-1) with the saponin-resistant fraction in NRK cells. NRK cells were treated with saponin, removed with a stream of PBS and examined for the presence of p52(PAI-1) by immunofluoresence (upper panel) or using one- and two-dimensional electrophoresis (lower panel). p52(PAI-1) is distributed at areas of cell attachment (rather than throughout the entire culture dish). Electrophoretic analysis of metabolically labeled de novo synthesized proteins indicates that p52(PAI-1) is a major constituent of this SAP fraction presenting as a complex of glycosylated species ranging from 43 to 52 kDa (arrow heads). While the 52 kDa glycosylated isoforms represent the predominant species, lower molecular weight isoforms are detected in the SAP fraction compartment that are immunoreactive with p52(PAI-1) antibodies (Higgins and Ryan, 1992). Actin (a) is also represented in this fraction.

protein synthesis (Ryan and Higgins, 1993). These observations suggest that p52(PAI-1) expression is an important element in the serum-induced growth response since members of the immediate early gene family are thought to regulate early events of cell cycle progression after growth activation (Ryan and Higgins, 1993; Lau and Nathans, 1987). The increase and subsequent decline in p52(PAI-1) expression during the time course of serum-stimulated growth activation suggests that its substrate deposition may influence proteolytic activity at early points in the proliferative response and, thereby, alter the stability or half-life of cell-to-substrate adhesive structures during this growth state transition. Indeed, under conditions of quiescence (serum starved), NRK cells are typically

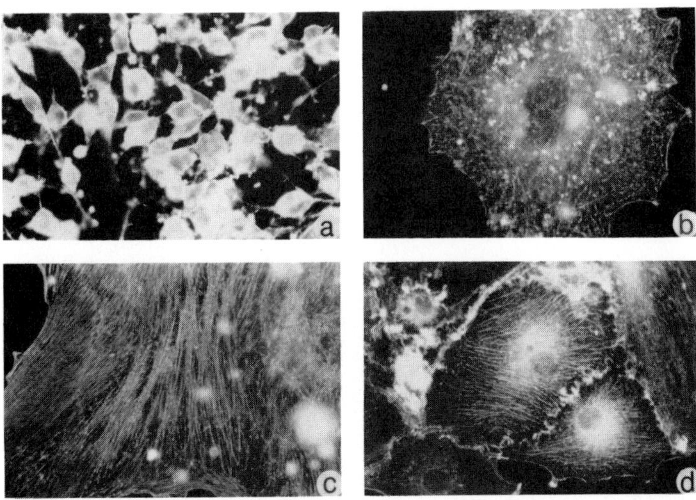

Figure 5. NaB-induced redistribution of F-actin-containing microfilaments in KNRK cells. KNRK cells (a) were exposed for 2 days (b), 3 days (c), or 74 days (d) to growth media containing 2 mM NaB and actin visualized with rhodamine-conjugated phalloidin on formalin fixed, NP-40-permeabilized cells. In contrast to KNRK cells, NaB-treated cells have a flattened morphology, possess fine actin filaments (evident within 2 days) and form fine parallel actin bundles (after long term exposure to NaB) (from Ryan and Higgins, 1988).

devoid of thick transcytoplasmic actin filaments and fail to display vinculin-associated focal contacts. Addition of 20% serum-containing medium initiates dramatic cytostructural changes indicative of a "stimulated" phenotype including the rapid formation of numerous actin stress fibers and vinculin-positive structures (Figure 10). Time course analysis indicates that creation of this stimulated phenotype occurred prior to induction and substrate deposition of p52(PAI-1) supporting the contention that accumulation of p52(PAI-1) in the region of the ventral undersurface (Figure 2) would most likely influence the stability rather than the initial formation of focal contacts (Ryan and Higgins, 1993). While the increased incidence of cell containing thick actin stress fibers are evident for up to 5 hours post-serum addition, the percentage of cells within the culture maintaining this stimulated phenotype subsequently decreases. These phenotypic changes coincide with decreased p52(PAI-1) mRNA abundance, protein biosynthesis and substrate deposition further suggesting that p52(PAI-1) influences these structures (Ryan and Higgins, 1993).

While the role of serum-mediated induced p52(PAI-1) expression in the growth response is not clear, it is tempting to speculate that p52(PAI-1) may influence the proliferative state of cells through the regulation of cell-to-ECM adhesion. In addition to specific growth factors, both the ECM and adhesive events likely influence cellular proliferative activity (Ingber, 1991). Regulation of substrate adhesion by p52(PAI-1) (via control of the stability of newly-formed focal contacts) may in turn modulate adhesion-based transduction pathways since occupancy of integrin receptors can mediate reprogramming of gene expression (Werb et al., 1989; Larjava et al., 1993).

Cytochalasin D-Mediated p52(PAI-1) Induction: Cell Shape-Associated Activation of p52(PAI-1) Gene Expression

The signal transduction pathway regulating p52(PAI-1) expression is controlled, in part, by specific growth factors although numerous genes (including the immediate-early response complement) are also activated through cell shape/adhesion mediated mechanisms (Dike and Farmer, 1988). The available data on both the K/NaB and serum-stimulated

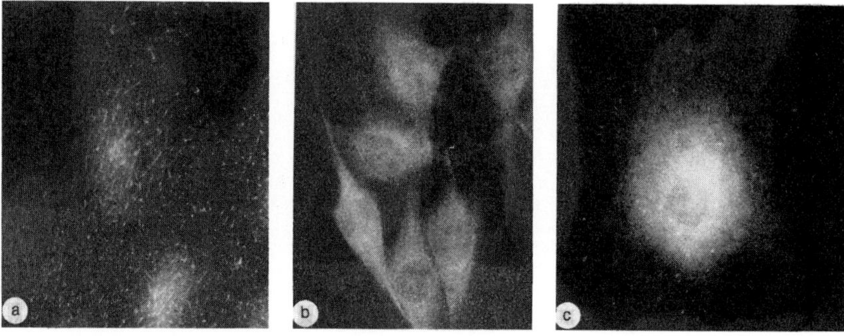

Figure 6. Formation of vinculin-containing adhesion plaques in K/NaB cells. Monoclonal anti-vinculin antibodies were used to visualize vinculin in NRK (a), KNRK (b), and NaB-treated KNRK (c) cells. While NRK cells contain numerous vinculin-containing adhesion plaques, KNRK cells are devoid of these structures. Addition of NaB to KNRK cells results in an increase incidence of vinculin positive structures at the cell periphery that are typical of focal contacts .

NRK cell systems clearly indicate that changes in cell morphology and adhesion (in particular those involving the actin cytoskeleton) were associated with subsequent reprogramming of p52(PAI-1) expression. To further examine the relationship between cell morphology and p52(PAI-1) gene expression, NRK cells were treated with the actin microfilament-disrupting agent, cytochalasin D (CD), to initiate dramatic cytoarchitectural alterations and the effect of such shape perturbations on p52(PAI-1) expression measured. CD disrupted actin filament organization and caused cell rounding in a concentration dependent manner with 90% of the cells affected by CD concentrations greater than 8 uM (Figure 11). This phenotypic response was accompanied by elevations in SAP fraction-associated p52(PAI-1) protein; such induction paralleled the increased incidence of rounded cells (Higgins et al., 1992). Increased SAP-associated p52(PAI-1) occurred within 2 hours of CD exposure was maximal at 4 hours (to levels 15-22 fold over control cultures) and subsequently declined thereafter (Higgins et al., 1992). Increased substrate-associated p52(PAI-1) accumulation in response to CD, as was the case for NaB-induction of p52(PAI-1) in KNRK cells, required on-going RNA synthesis during the period of inducer treatment (Figure 12) and coincided with an induction of mRNA at both the steady-state and transcriptional levels (Figure 13).

Since cell shape and growth factor responsiveness may be interrelated, it was necessary to determine if CD could act as an inducer of p52(PAI-1) expression in the absence of exogenously added serum growth factors since previous assessments were made in 10% serum-containing media (Higgins, et al., 1989b, 1990). NRK cells were exposed to CD in 10% FBS or under completely serum-free conditions. Even in the absence of serum, CD

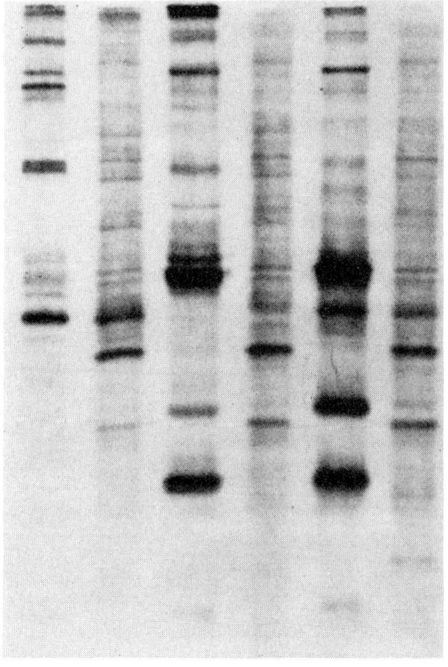

Figure 7. NaB induces the secretion and substrate accumulation of p52(PAI-1) in NaB-treated KNRK cells. NRK, KNRK and NaB-treated (for 3 days) cultures were metabolically labeled with ^{35}S-methionine and the secreted (SP) and substrate associated material (SAM) separated on 10%-SDS gels. p52(PAI-1) [p52] is present in both the SP and SAM fractions of NRK K/NaB cultures. p52(PAI-1) is clearly absent, however, from both of these fractions in KNRK (from Higgins and Ryan, 1989a).

was an effective inducer of p52(PAI-1) biosynthesis (Higgins et al., 1992). To ensure the effectiveness of CD as an inducer under growth factor deprived conditions, NRK cells were made quiescent by serum deprivation prior to being exposed to CD in serum free media. Induction was still evident (at levels comparable to conditions containing serum) demonstrating that CD can act as a complete inducer of p52(PAI-1), lending support that p52(PAI-1) expression is regulated in a cell shape dependent manner.

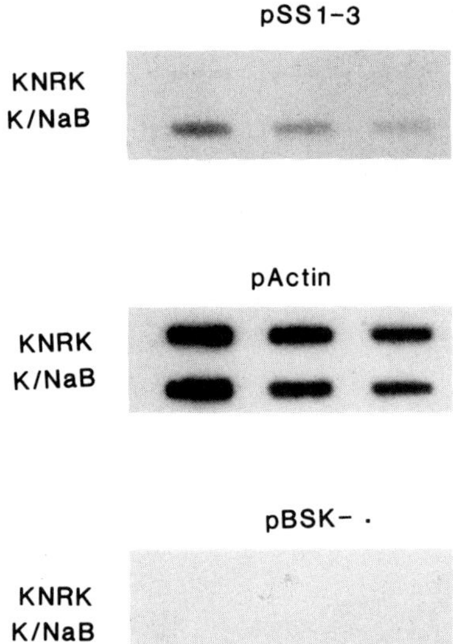

Figure 8. NaB-mediated induction of p52(PAI-1) is associated with an increase in steady state levels of p52(PAI-1) mRNA. Total cytoplasmic RNA (20, 10, 5 ug/slot) was blotted onto nitrocellulose and hybridized with ^{32}P-labeled cDNAs to p52(PAI-1) [pSS1-3] and actin. Cloning vector alone (pBSK) was used as a control probe for non-specific hybridization. Induced p52(PAI-1) expression in K/NaB cultures is evident upon examination of p52(PAI-1) mRNA. Compared to KNRK, K/NaB cultures contain substantially greater levels of p52(PAI-1) cytoplasmic mRNA. Actin mRNA abundance, however, does not change (from Higgins and Ryan, 1991).

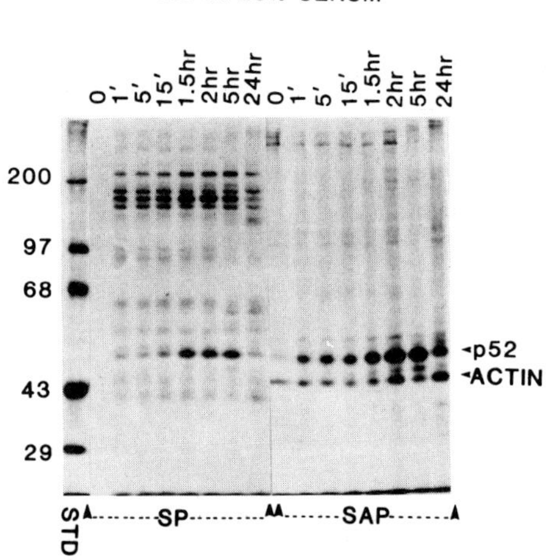

Figure 9. Induction of p52(PAI-1) expression following growth stimulation of quiescent NRK cells. NRK cells were maintained in 0.5% serum-containing medium for 3 days to achieve quiescence prior to being growth stimulated with medium containing 20% serum for times indicated. Cultures were subsequently labeled with ^{35}S-methionine, the secreted protein (SP) and saponin-resistant (SAP) fractions collected and equal cpm of TCA-insoluble proteins separated on 10%-SDS gels. Exposure of quiescent NRK cells (time 0) to serum induced the rapid accumulation of p52(PAI-1) in both the SP and SAP fractions. p52(PAI-1), actin and labeled protein standards are indicated (from Rayn and Higgins, 1993).

DISCUSSION

NRK cells are a useful model system to examine the relationship between p52(PAI-1) expression and cytoarchitecture. Events leading to changes in p52(PAI-1) expression are quite variable with respect to inducers (i.e., NaB, serum, CD). A commonality to each of these, however, is that p52(PAI-1) expression appears closely linked to the reorganization of the actin cytoskeleton and the cell-to-ECM adhesion apparatus. Based on these data, we have suggested that p52(PAI-1) is an important regulator of cell shape and substrate adhesion through controlled pericelluar proteolysis at focal adhesion points. Physiologic and non-physiologic processes associated with morphological changes in NRK cells may, therefore, influence or be influenced by changes in p52(PAI-1) expression.

Regulation of cell shape and adhesion by p52(PAI-1) could potentially influence cellular

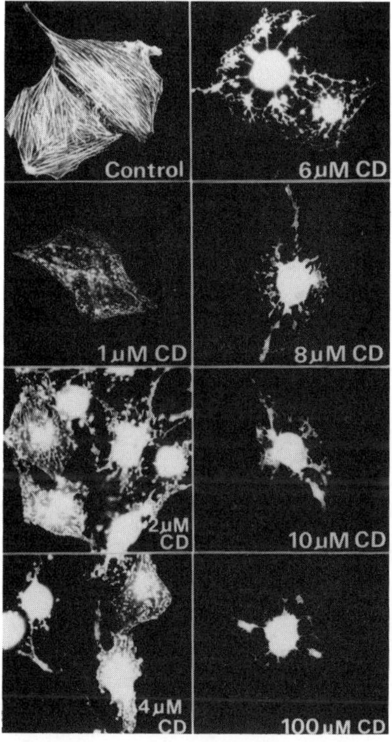

Figure 10. Serum-mediated growth stimulation of quiescent NRK cells results in a rapid reorganization of the actin cytoskeleton and formation of vinculin-containing adhesion plaques. Quiescent NRK cells were growth stimulated (for times indicated); actin microfilaments and focal contacts were visualized with rhodamine-conjugated phalloidin and anti-vinculin antibodies, respectively. Vinculin-positive structures are absent from quiescent NRK cells while those exposed to serum (2 hr.) contain numerous vinculin-containing structures indicative of focal contacts. Serum stimulation is also associated a dramatic increase in cells containing thick, transcytoplasmic microfilaments (from Ryan and Higgins, 1993).

growth since morphology is an important aspect of proliferative control (Folkman and Moscana, 1978; Ben-Ze'ev, 1980, 1986; Watt, 1986; Bissell and Barcellos-Hoff, 1987; McDonald, 1989; Ingber, 1990, 1991; Ingber et al., 1990; Bellas et al., 1991). That progression through the cell cycle is governed by cell shape/adhesion is supported by evidence that genes encoding cytoskeletal (actin, vinculin, tropomyosin), cell-to-substrate-associated (integrin), and ECM (fibronectin) proteins are also represented in the immediate early response gene family (Ryseck et al., 1989; Bellas et al., 1991). Similarly, proteins involved in pericellular proteolysis (u-PA, PAI-1) and which are localized to areas of substrate adhesion, are members of this family. Typical of these genes is a rapid induction without the need of _de novo_ protein synthesis, followed by a decrease (in some instances), to an undetectable level prior to entry into the DNA synthetic phase of the cell cycle (Lau and Nathans, 1987; Herschman, 1991). In the case of PAI-1 and u-PA, this type of regulated expression would be expected to influence stability of focal contacts formed early

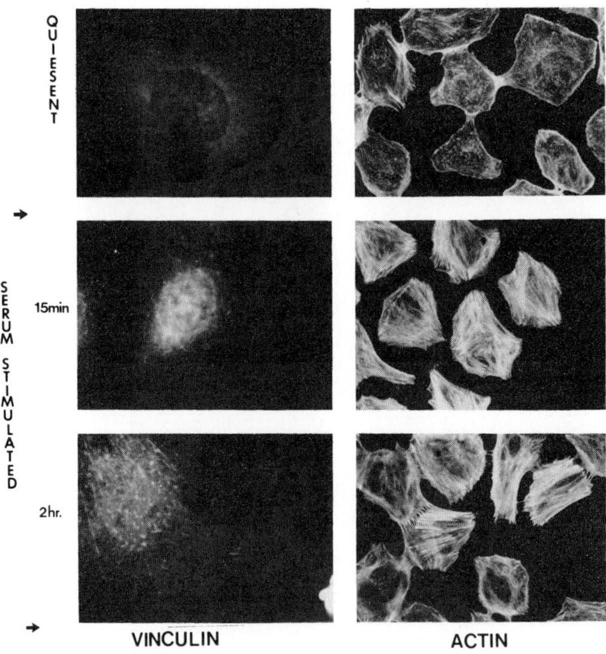

Figure 11: Cytochalasin D-mediated microfilament disruption and cell shape changes in NRK cells is concentration dependent. NRK cells were treated with growth medium containing various concentrations of cytochalasin D (CD) and the actin microfilament network visualized with rhodamine-conjugated phalloidin. Low concentrations of CD (1 uM) was effective at disrupting actin filaments without cell rounding, however, concentrations greater than 2 uM resulted in an increase in the percentage of cells having a round phenotype. Maximal morphological changes were evident at $\geq$8 uM (from Higgins et al., 1992).

after growth stimulation (primarily in the G1 stage) further implicating these changes as an important constituent of cell growth control.

The relevance of focal contact formation in either cellular growth response or gene re-programming is unclear. Increasing evidence suggests that an integrin-mediated signaling pathway is initiated after cell-to-substrate adhesion or integrin receptor clustering involving, perhaps, the p125 focal adhesion kinase (p125[FAK]) which is localized at the cytoplasmic face of the focal contact (Burridge, 1992; Hanks, 1992; Kornberg et al., 1992;

Lipfert et al., 1992; Hildebrand et al., 1993). Activation of p125[FAK], in response to receptor clustering, appears to be associated with phosphorylation of specific focal contact-associated proteins which, in turn, may be important in altering the nature of these structures and/or in subsequent activation of gene expression (Burridge et al., 1992). Induction of immediate-early response genes through intergin-mediated adhesion highlights the potential importance of adhesion induced signaling (Dike and Farmer, 1988, 1989). Whether integrin-dependent events in general or p125[FAK] in particular is involved in

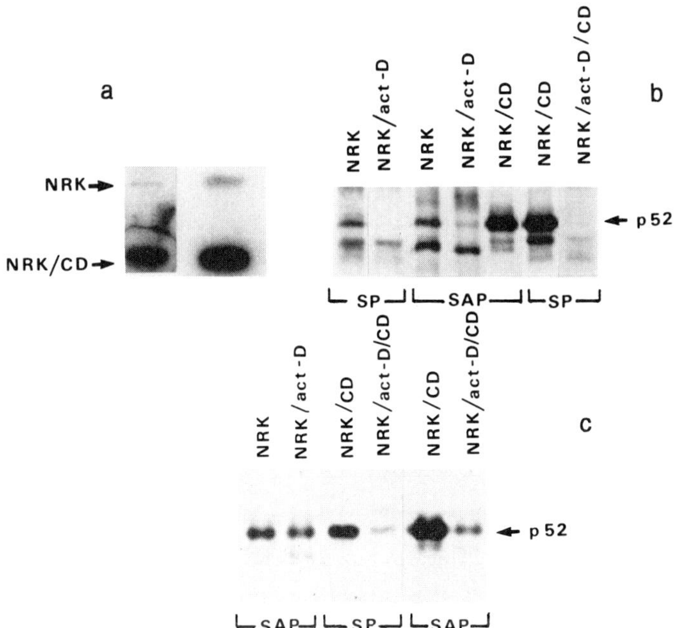

Figure 12. CD-mediated p52(PAI-1) RNA and protein induction in NRK cells. NRK cells were treated with 100 uM CD prior to analysis of p52(PAI-1) RNA and protein content. Cytoplasmic RNA (20 ug) from NRK/CD cultures were blotted to nitrocellulose and hybridized with a p52(PAI-1) cDNA probe (a). Alternatively, cultures were exposed to CD in the presence of the RNA synthesis inhibitor, actinomycin D (act-D) at 0.16 (b) or 6 uM (c) for a total of 24 and 4 hours respectively, prior to metabolic labeling with ^{35}S-methionine and isolation of the SP and SAP fractions. Equal TCA-insoluble cpm of protein were analyzed on 10%-SDS gels. Exposure of NRK cells to CD resulted in a 10- to 12-fold increase in p52(PAI-1) mRNA (a). p52(PAI-1) accumulation into both the SP and SAP fractions was blocked by act-D (b and c), demonstrating that CD-mediated p52(PAI-1) induction in NRK cells requires RNA synthesis (modified from Higgins et al., 1990).

p52(PAI-1) expression control remains to be determined. Studies with CD, however, suggest that p52(PAI-1) expression is at least influenced by cell shape and adhesion; integrin- or p125[FAK]-dependent pathways are primary candidates in this signal transduction pathway. Regardless of the inductive mechanism(s), changes in p52(PAI-1) expression could be an important element in cell shape-dependent transduction pathways since pericellular proteolytic activity influences the focal contact and potentially signaling mechanisms associated with these adhesive structures.

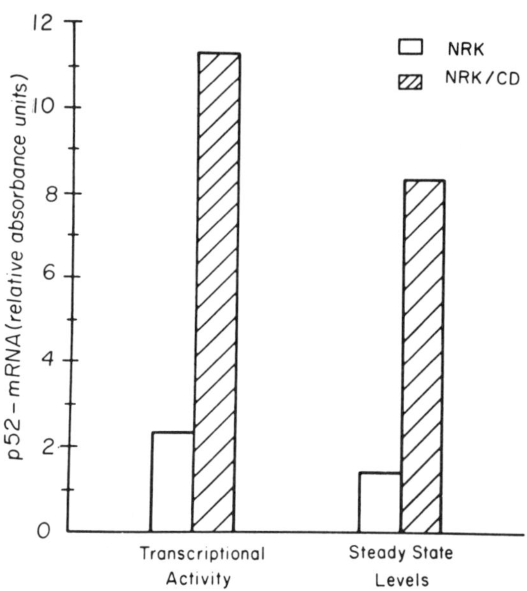

Figure 13. Transcriptional activation of the p52(PAI-1) gene by cytochalasin D in NRK cells. Newly transcribed nuclear and steady state levels of cytoplasmic RNA were measured from NRK (open bars) and CD-treated (for 24 hours) NRK cells (hatched bars). For measurement of transcriptional activity, nuclei were isolated, transcription of nascent RNAs continued in vitro in the presence of ^{32}P-UTP and the labeled RNAs hybridized to p52(PAI-1) cDNA immobilized to nitrocellulose. For steady state measurements, cytoplasmic RNA was separated on agarose/formaldehyde gels, transferred to nylon membranes and hybridized with a labeled p52(PAI-1) probe. Signal intensity was measured by densitometry and expressed as arbitrary absorbance units (from Higgins et al., 1992).

Acknowledgments

This work was supported by grants from the National Institutes of Health (GM4261, DK46272), from the National Dairy Board (administered by the National Dairy Council), and from Boehringer Mannheim Italia.

REFERENCES

Altenburg, B.C., Via, D.P., and Steiner, S.H., 1976, Modification of the phenotype of murine sarcoma virus-transformed cells by sodium butyrate, *Exp. Cell Res.* 102:223.

Bellas, R.E., Bendori, F., and Farmer, S.R., 1991, Epidermal growth factor activation of vinculin and β_1-integrin gene transcription in quiescent 3T3 cells, *J. Biol. Chem.* 266:12008.

Ben-Ze'ev, A., 1980, Protein synthesis requires cell-surface contact while nuclear events respond to cell shape in anchorage-dependent fibroblasts, *Cell.* 21:365.

Ben-Ze'ev, A., 1986, The relationship between cytoplasmic organization, gene expression and morphogenesis, *TIBS.* 11:478

Ben-Ze'ev, A., 1987, The role of changes in cell shape and contacts in the regulation of cytoskeleton expression during differentiation, *J. Cell Sci. Suppl.* 8:293-312.

Bissell, M.J., and Barcellos-Hoff, M.H.J., 1987, The influence of extracellular matrix on gene expression: is structure the message?, *J. Cell Sci, Supp.* 8:327.

Bockus, B.J. and Stiles, C.D., 1984, Regulation of cytoskeletal architecture by platelet-derived growth factor, insulin and epidermal growth factor, *Exp. Cell Res.* 153:186.

Burridge, K., Turner, C.E., and Romer, L.H.,1992, Tyrosine phosphorylation of paxillin and pp125[FAK] accompanies cell adhesion to extracellular matrix: a role in cytoskeletal assembly, *J. Cell Biol.* 119:893.

Carley, W.W., and Webb, W.W., 1983, F-actin aggregates may activate transformed cell surfaces, *Cell Motil.* 3:383.

Ciambrone, G.J. and McKeown-Longo, P.J., 1990, Plasminogen activator inhibitor type-1 stabilizes vitronectin-dependent adhesions in HT1080 cells, *J. Cell Biol.* 111:2183.

Cooper, H.L., Feuerstein, N., Noda, M., and Bassin, R.H., 1985, Suppression of tropomyosin synthesis, a common biochemical feature of oncogenesis by structurally diverse retroviral oncogenes, *J. Cell Biol.* 5:972.

Cooper, J.A.,1987, Effects of cytochalasin and phalloidin on actin, *J. Cell Biol.* 105:1473.

Dalton, S.L., Marcantonio, E.E., and Assoian, R.K.,1992, Cell attachment controls fibronectin and $\alpha 5\beta 1$ integrin levels in fibroblasts, *J. Biol. Chem.* 267:8186.

Dano, K., Behrendt, N., Lund, L.R., Ronne, E., Pollanen, J., Salonen, E.M., Stephens, R.W., Tapiovaara, H., and Vaheri, A., 1989,"Cancer Metastisis. Molecular and Cellular Biology, Host Immune Responses and Perspectives for Treatment", V. Schirrmacher and R. Schwarz-Albiez, eds., Springer,Verlag, Berlin.

Dike, L.E. and Farmer, S.R.,1988, Cell adhesion induces expression of growth-associated genes in suspension-arrested fibroblasts, *Proc. Natl. Acad. Sci. USA.* 85:6792.

Ezzell, C., 1993, Just the fak(s) ma'am: Researchers investigate a new signalling enzyme, J. NIH Res., 5:49.

Farmer, S.R. and Dike, L.E.,1989, Cell shape and growth control:role of cytoskeleton-extracellular matrix interactions, *in*: "Cell shape determinants; Regulation and Regulatory Role", W.D. Stein and F. Bronner, eds., Academic Press, New York, New York.

Fernandez, J.L.R. and Ben-Ze'ev, A., 1989 Regulation of fibronectin, integrin and cytoskeleton expression in differentiating adipocytes: inhibition by extracellular matrix and polylysine, *Differentiation.* 42:65.

Folkman, J., and Moscana, A., 1978, Role of cell shape in growth control, *Nature.*273:345.

Goodman, G. and Miranda, A.,1978, "Cytochalasins-Biochemical and Cell Biology Aspects", S. Tannenbaum, ed., Elsevier, Amsterdam.

Hayman, E.G., Engvall,E., and Ruoslahti, E., 1981, Concomitant loss of cell surface fibronectin and laminin from transformed rat kidney cells, *J. Cell Biol.* 88:352.

Herman, B. and Pledger, W.J.,1985, Platelet-derived growth factor-induced alteration in vinculin and actin distribution in BALB/c-3T3 cells, *J. Cell Biol.* 100:1013.

Herschman, H.R., 1991, Primary response genes induced by growth factors and tumor promoters, *Ann. Rev Biochem.* 173:93.

Higgins, P.J. and Ryan, M.P., 1989a, Biochemical localization of the transfromation-sensitive 52 kDa (p52) protein to th substratum contact regions of cultured rat fibroblasts, *Biochem J.*257:173.

Higgins, P.J., Ryan, M.P., and Chaudhari, P., 1989b, Cytochalasin D-mediated hyperinduction of the substrate-associated 52-kilodalton protein p52 in rat kidney fibroblasts, *J. Cell. Physiol.* 139:407.

Higgins, P.J., Ryan, M.P., Zehb, R., Gelehrter, T.D., and Chaudhari, P., 1990, p52 induction by cytochalasin D in rat kidney fibroblasts: homologies between p52 and plasminogen activator inhibitor type-1, *J. Cell. Physiol.* 143:321.

Higgins, P.J., Chaudhari, P., and Ryan, M.P., 1991, Cell-shape regulation and matrix protein p52 content in phenotypic variants of *ras*-transformed rat kidney fibroblasts, *Biochem. J.* 273:651.

Higgins, P.J. and Ryan, M.P., 1991, p52(PAI-1) and actin expression in butyrate-induced flat revertants of ras-transformed rat kidney cells, *Biochem. J.* 279:883.

Higgins, P.J., Ryan, M.P., and Ahmed, A., 1992, Cell-shape-associated transcriptional activation of the p52(PAI-1) gene in rat kidney cells, *Biochem. J.* 288:1017.

Higgins, P.J. and Ryan, M.P., 1992, Identification of the 52 kDa cytoskeletal-like protein of cytochalasin D-stimulated normal rat kidney (NRK/CD) cells as substrate-associated glycoprotein p52 [plasminogenactivator inhibitor type-1 (PAI-1)], *Biochem. J.* 284:433.

Hildebrand, J.D., Schaller, M.D., and Parsons, J.T.,1993, Identification of sequences required for the efficient localization of the focal adhesion kinase, pp125[FAK], to cellular focal adhesions, *J. Cell Biol.* 123:993.

Ingber, D., 1990, Fibronectin controls capillary endothelial cell growth by modulating cell shape, *Proc. Natl. Acad. Sci. USA.* 87:3579.

Ingber, D., Prusty, D., Frangioni, J.V., Cragoe, E.J., Lechene, C., and Schwartz, M.A., 1990, Control of intracellular pH and growth by fibronectin in capillary endothelial cells, *J. Cell Biol.* 110:1803.

Ingber, D., 1991, Extracellular matrix and cell shape: potential control points for inhibition of angiogenesis, *J. Cell. Biochem.* 47:236.

Laiho, M. and Keski-Oja, J., 1989, Growth factors in the regulation of pericellular proteolysis, *Can. Res.* 49:2522.

Larjava, H., Lyons, J.G., Salo,T., Makela, M., Koivisto, L., Birkedal-Hansen, H., Akiyama, S.K., Yamada, K., and Heino, J., 1993, Anti-integrin antibodies induce type IV collagenase expression in keratinocytes, *J. Cell. Physiol.* 157:190.

Lau, L.F., and Nathans, D., 1987, Expression of a set of growth-regulated immediate early genes in BALB/c 3T3 cells: coordinate regulation with c-*fos* or c-*myc*. *Proc. Natl. Acad. Sci. USA.* 84:1182.

Lipfert, L., Haimovich, B., Schaller, M.D., Cobb, B.S., Parsons, J.T., and Brugge, J.S., 1992, Integrin-dependent phosphorylation and activation of the protein tyrosine kinase pp125^{FAK} in platelets, *J. Cell Biol.* 119:905.

Maness, P.F.,1981, Actin structure in fibroblasts-its possible role in transfromation and tumorigenesis, *in*: "Cell and Muscle Motility", R.M. Dowben and J.W. Shaw, eds., Plenum Publishing, New York.

Matsumura, F., Lin, J.J.-C., Yamashiro-Matsumura, S., Thomas, G.P., and Topp, W.C., 1983, Differential expression of tropomyosin forms in the microfilaments isolated from normal and transformed rat cultured cells, *J. Biol. Chem.* 258:13594.

McDonald, J.A., 1989, Matrix regualtion of cell shape and gene expression, *Curr. Opin. Cell Biol.* 1:995.

Pollanen, J., Saksela, O., Salonen, E-M., Andreasen, P., Neilsen, L., Dano, K., and Vaheri, A., 1987, Distinct localization of urokinase-type plasminogen activator and its type-1 inhibitor under cultured human fibroblasts and sarcoma cells, *J.Cell Biol.* 104:1085.

Pollanen, J., Stephens, R.W., and Vaheri, A., 1991, Directed plasminogen activation at the surface of normal and malignant cells, *Adv. Can.* Res. 57:273-328.

Ridley, A. and Hall, A., 1992, The small GTP-binding protein rho regulates the assembly of focal adhesions and actin stress fibers in response to growth factors, *Cell.* 70:389.

Ryseck, R.P., MacDonald-Bravo, H., Zerial, M., and Bravo, R., 1989, Coordinate induction of fibronectin, fibronectin receptor, tropomyosins, and actin genes in serum-stimulated fibroblasts, *Exp. Cell Res.* 180:537.

Ryan, M.P. and Higgins, P.J., 1988, Cytoarchitecture of kirsten sarcoma virus-transformed rat kidney fibriblasts: butyrate-induced reorganization within the actin microfilament network, *J. Cell. Physiol.* 137:25.

Ryan, M.P. and Higgins, P.J., 1989, Sodium-n-butyrate induces secretion and substrate accumulation of p52 in kirsten sarcoma virus-transformed rat kidney fibroblasts, *Int. J. Biochem.* 21:31.

Ryan, M.P. and Higgins, P.J., 1993, Growth state-regulated expression of p52(PAI-1) in normal rat kidney cells, *J. Cell. Physiol.* 155:376.

Schliwa, M., 1982, Action of cytochalasin D on cytoskeletal networks, *J. Cell Biol.* 92:79-81.

Spiegelman, B.M. and Farmer, S.R., 1982, Decrease in tubulin and actin gene expression prior to morphological differentiation of 3T3 adipocytes, Cell. 29:53

Stickel, S.K., and Wang, Y-l.,1987, Alpha-actinin-containing aggregates in transformed cells are highly dynamic structures, *J. Cell Biol.* 104:1521

Wang, Y-l., and Golgberg, A.R.,1976, Changes in microfilament organization and surface topography upon transformation of chick embryo fibroblasts with Rous sarcoma virus, *Proc. Natl. Acad. Sci. USA.* 73:4056.

Watt, F.M., 1986, The extracellular matrix and cell shape, *TIBS.* 11:482.

Werb, Z., Tremble, P.M., Behrendtsen, O., Crowley, E., and Damsky, C.H., 1989, Signal transduction through the fibronectin receptor induces collagenase and stromelysin gene expression, *J. Cell Biol.* 109:877.

Wahrman, M.Z., Gagnier, S.E., Kobrin, D.R., Higgins, P.J., and Augenlicht, L.H., 1985, Cellular and molecular changes in 3T3 cell transformed spontaneously or by DNA transfection, *Tumour Biol.* 6:41.

Contributors

Klaus Aktories
Institute of Pharmacology and Toxicology
University of the Saarland
D 4630 Bochum, Germany

Gary J. Bassell
Department of Cell Biology
University of Massachusetts Medical School
Worcester, MA 01655

Daniel ben-Avraham
Physics Department
Clarkson University
Potsdam, NY 13699-5820

Avri Ben-Ze'ev
Department of Molecular Genetics & Virology
Weizmann Institute of Science
Rehovot 76100, Israel

Marie-France Carlier
Laboratoire d'Enzymologie
C.N.R.S., Gif-sur-Yvette
Cedex, France

Cecile Combeau
Laboratoire d'Enzymologie
C.N.R.S., Gif-sur-Yvette
Cedex, France

Jimmy H. Collins
Department of Biochemistry
Temple University School of Medicine
Philadelphia, PA 19140

Lawrence E. Crawford
Department of Medicine
Johns Hopkins University School of Medicine
Baltimore, MD 21287

Jyotsna Dhawan
Department of Biochemistry
Boston University School of Medicine
Boston, MA 02118

Andrea Ditsch
Institute of Physiological Chemistry
Ruhr-University
D 4630 Bochum, Germany

Cris G. dos Remedios
Muscle Research Unit
University of Sydney
NSW 2006, Australia

Anh M. Duong
Chemistry & Biochemistry
University of California
Los Angeles, CA 90024

James E. Estes
Research Service
Stratton VA Medical Center
Albany, NY 12208

Stephen R. Farmer
Department of Biochemistry
Boston University School of Medicine
Boston, MA 02118

Stephane Fievez
Laboratoire d'Enzymologie
C.N.R.S., Gif-sur-Yvette
Cedex, France

Benjamin Geiger
Chemical Immunology
Weizmann Institute of Science
Rehovot 76100, Israel

Lewis C. Gershman
Research and Medical Services
Stratton VA Medical Center
Albany, NY 12208

Ursula Gluck
Department of Molecular Genetics & Virology
Weizmann Institute of Science
Rehovot 76100, Israel

Pascal J. Goldschmidt-Clermont
Cell Biology, Anatomy and Medicine
Johns Hopkins University School of Medicine
Baltimore, MD 21287

Enrico Grazi
Istituto di Chimica Biologica
Universita di Ferrara
Ferrara, Italy

Brett D. Hambly
Department of Pathology
University of Sydney
NSW 2006, Australia

Alan W. Heldman
Department of Medicine
Johns Hopkins University School of Medicine
Baltimore, MD 21287

Paul J. Higgins
Microbiology, Immunology & Molecular Genetics
Albany Medical College
Albany, NY 12208

Sarah E. Hitchcock-DeGregori
Department of Neuroscience & Cell Biology
Robert Wood Johnson Medical School
Piscataway, NJ 08854

Kenneth C. Holmes
Max-Planck-Institute
6900 Heidelberg
Germany

Ingo Just
Institute of Pharmacology & Toxicology
University of the Saarland
Homburg/Saar, Germany

Peter Kei*B***ling**
Muscle Research Unit
University of Sydney
NSW 2006, Australia

Henry J. Kinosian
Research & Medical Services
Stratton VA Medical Center
Albany, NY 12208

Edward H. Kislauskis
Department of Cell Biology
University of Massachusetts
Worcester, MA 01655

Fumio Matsumura
Molecular Biology & Biochemistry
Rutgers University
Piscataway, NJ 08855-1059

Robert Mendelson
Biochemistry & Biophysics
University of California
San Francisco, CA 94143

Edward Morris
Biophysics Section
Blackett Laboratory
Imperial College, London

Dominique Pantaloni
Laboratoire d'Enzymologie
C.N.R.S., Gif-sur-Yvette
Cedex, France

Christine M. Powers
Department of Cell Biology
University of Massachusetts
Worcester, MA 01655

Ayyappan K. Rajasekaran
Cell Biology & Anatomy
Cornell University Medical College
New York, NY 10021

Emil Reisler
Chemistry & Biochemistry
University of California
Los Angeles, CA 90024

Jose Luis Rodriguez-Fernandez
Department of Molecular Genetics & Virology
Weizmann Institute of Science
Rehovot 76100, Israel

Anthony Ross
Department of Cell Biology
University of Massachusetts Medical School
Worcester, MA 01655

Michael P. Ryan
Microbiology, Immunology & Molecular Genetics
Albany Medical College
Albany, NY 12208

Daniela Salomon
Chemical Immunology
Weizmann Institute of Science
Rehovot 76100, Israel

Beate Schoepper
Institute of Physiological Chemistry
Ruhr-University
D 4630 Bochum, Germany

Suzanne B. Schwartz
Department of Surgery
Cornell University Medical College
New York, NY 10021

Lynn A. Selden
Research & Medical Services
Stratton VA Medical Center
Albany, NY 12208

Norma Selve
Institute of Physiological Chemistry
Ruhr-University
D 4630 Bochum, Germany

Robert H. Singer
Department of Cell Biology
University of Massachusetts Medical Center
Worcester, MA 01655

Lisa Staiano-Coico
Department of Surgery
Cornell University Medical College
New York, NY 10021

Cindi L. Sundell
Department of Cell Biology
University of Massachusetts
Worcester, MA 01655

Helena Swanljung-Collins
Department of Biochemistry
Temple University Medical School
Philadelphia, PA 19140

Krishan L. Taneja
Department of Cell Biology
University of Massachusetts
Worcester, MA 01655

Julie A. Theriot
Biochemistry & Biophysics
University of California
San Francisco, CA 94143-0448

Monique M. Tirion
Physics Department
Clarkson University
Potsdam, NY 13699-5820

Robert W. Tucker
Cell Biology, Anatomy & Oncology
Johns Hopkins University
Baltimore, MD 21287

Catherine Valentin-Ranc
Laboratoire d'Enzymologie
C.N.R.S., Gif-sur-Yvette
Cedex, France

Albrecht Wegner
Institute of Physiological Chemistry
Ruhr-University
D 4630 Bochum, Germany

Michaela Wille
Pharmacology & Toxicology Institute
University of the Saarland
Homburg/Saar, Germany

Yoshihiko Yamakita
Molecular Biology & Biochemistry
Rutgers University
Piscataway, NJ 08855-1059

Shigeko Yamashiro
Molecular Biology & Biochemistry
Rutgers University
Piscataway, NJ 08855-1059

Kyonsoo Yoshida
Molecular Biology & Biochemistry
Rutgers University
Piscataway, NJ 08855-1059